ADVANCED COURSES OF MATHEMATICAL ANALYSIS III

EDITORS

JUAN M. DELGADO SÁNCHEZ
University of Huelva, Spain

TOMÁS DOMÍNGUEZ BENAVIDES
University of Seville, Spain

PROCEEDINGS OF THE THIRD INTERNATIONAL SCHOOL

ADVANCED COURSES OF MATHEMATICAL ANALYSIS III

La Rábida, Spain 3 – 7 September 2007

World Scientific

NEW JERSEY • LONDON • SINGAPORE • BEIJING • SHANGHAI • HONG KONG • TAIPEI • CHENNAI

Published by

World Scientific Publishing Co. Pte. Ltd.

5 Toh Tuck Link, Singapore 596224

USA office: 27 Warren Street, Suite 401-402, Hackensack, NJ 07601

UK office: 57 Shelton Street, Covent Garden, London WC2H 9HE

British Library Cataloguing-in-Publication Data
A catalogue record for this book is available from the British Library.

ADVANCED COURSES OF MATHEMATICAL ANALYSIS III
Proceedings of the Third International School

ISBN-13 978-981-281-844-7
ISBN-10 981-281-844-8

Printed in Singapore.

And, though the warrior's sun has set,
Its light shall linger round us yet,
Bright, radiant, blest.

(J. Manrique; H. W. Longfellow)

To Antonio Aizpuru

In Memoriam

PREFACE

The III International Course of Mathematical Analysis in Andalusia, held in La Rábida (Huelva), 3–7 September 2007, continued the tradition of previous courses in Cádiz (2002) and Granada (2004). Five years ago, representatives of several Andalusian universities made a concerted effort organizing a course to provide an extensive overview of the research in different areas of Mathematical Analysis. The friendly cooperation of many Andalusian research groups in these areas and the initiative of Antonio Aizpuru and Fernando León made possible the organization of the first course in Cádiz. A new and wider cooperation of the Andalusian research groups in Real Analysis, Complex Variable and Functional Analysis and, mainly, the encouragement and hard work of María Victoria Velasco were the cornerstone to support the second course in Granada. During the Gala Dinner in this course (held in a beautiful house which was used as a summer palace by the latter Arab–Andalusian Kings), a group of professors from the universities of Sevilla and Huelva agreed to organize the third course. With the support of the Spanish National Government, the universities of Seville and Huelva, sponsored by several private and official institutions and hosted by the International University of Andalusia (Sede Iberoamericana de La Rábida) where we could invite some leading researchers in this area to give three seminars and eleven plenary lectures. The course brought more than 70 participants from different countries and provided an ideal forum for learning and exchanging of ideas. The high scientific quality of the lectures and seminars offered in this course made us think about the interest of a book collecting these talks. We asked the speakers for a written version of their lectures. The lecturers kindly facilitated us and we could agree with World Scientific Publishing Co. the publication of these proceedings which can be of high interest to graduate students and researchers in several areas of Mathematical Analysis.

The present book includes the contributions corresponding to the seminars by Marco Abate, Eleonor Harboure and Edward Odell and to the plenary talks by Óscar Blasco, Joaquim Bruna, Bernardo Cascales, Francisco

L. Hernández, Lawrence Narici, Héctor Salas, Bertram Schreiber, Antonio Villar and Wiesław Żelazko. Seminars and talks lectured by these prestigious researchers attracted the interest of a big number of graduate students and researchers who attended this conference. The excellent work of the lecturers and the scientific contributions from them and all participants made possible the success of this course.

The talks of M. Abate concerns the theory of local discrete dynamical systems in complex dimension 1, describing what is known on the topological and dynamical structure of the stable set, and on topological, holomorphic and formal conjugacy classes.

O. Blasco presents DeLeeuw-type transference theorems for bilinear multipliers. They allow us to obtain the boundedness of the periodic and discrete versions of bilinear multipliers (even for their maximal versions) and to get new applications of these results in Ergodic Theory.

J. Bruna studies functions which generate the Lebesgue space by translations. He shows that the discrete translation parameter sets $\Lambda \subset \mathbb{R}$ for which some $\varphi \in L_1(\mathbb{R})$ exists such that the translates $\varphi(x-\lambda)$, $\lambda \in \Lambda$, span $L_1(\mathbb{R})$ are exactly the uniqueness sets for certain quasianalytic classes, and gives explicit constructions of such generators φ.

B. Cascales' lecture shows that several classical results about compactness in functional analysis can all be derived from some suitable inequalities involving distances to spaces of continuous or Baire one functions. In particular, he gives quantitative versions of Grothendieck's characterization of weak compactness in spaces $C(K)$, and Eberlein–Grothendieck and Krein–Smulyan theorems.

E. Harboure poses several situations in analysis where some kind of smooth functions play a fundamental role. In connection with the study of Laplace equation, she analyzes the behavior of the fractional integral operator on L_p-spaces and presents a brief description of Besov spaces and their connection with a problem of non-linear approximation of a function by its wavelet expansion.

The domination problem for positive operators between Banach function spaces consists in given two positive operators $0 \leq R \leq T$ between two Banach lattices E and F and assuming that T belongs to a certain operator class, should R belong to the same class? F. Hernández surveys recent results on the behavior of related operator classes like strictly singular (or *Kato*) operators, strictly co-singular (or *Pelczynski*) operators as well as their local versions.

L. Narici explains to us many facts concerning the Hanh–Banach Theorem and the significant role played by the Austrian mathematician Eduard Helly in the development of this theorem.

The contribution by E. Odell discusses the Banach space structure of $L_p[0, 1]$, mostly in the reflexive setting, $1 < p < \infty$. This classical Banach space has been a prime case study for abstraction to a more general study of Banach space structure. E. Odell revises the most relevant properties of this space, concerning complemented and non-complemented subspaces, embeddings, normalized unconditional sequences, distortions of the norm, etc.

H. Salas revises properties of hypercyclic operators and presents several problems, some of them in the new classes of dual hypercyclic operators and frequently hypercyclic operators.

In B. Schreiber's lecture, the Operator Algebra Basic Theory is outlined and some applications are described. Many of these applications lead easily to open problems worthy of investigation, both in the area of the application and in the development of the basic theory.

A. Villar's lecture deals with the mathematical tools that permits to understand the logics of competitive markets. It refers to the solvability of a finite system of equations with non-negativity restrictions. Some changes in the environment are also considered: non-competitive behavior, non-convex feasible sets, non-finite sets of markets, a continuum of agents, etc.

W. Żelazco discusses some recent results and some open problems concerning unital F-algebras (i.e., a topological algebra which is an F-space). The following questions are considered:

(1) When are all maximal ideals closed?
(2) When are all ideals closed?
(3) When does a dense principal ideal exist?

There were many other pluses to the course; the tourist trip across Huelva Ría on a typical boat with dinner on board, the memorable trip to the Riotinto Mine Park and to the Marvels's Cave in Sierra de Aracena Natural Park, not to mention the BBQ and Gala Dinner held in the garden of La Rábida Residence followed by an amazing "Cheen–Cheen-Poom" dancing.

No conference can succeed without a lot of generous support and we would like to express our gratitude to the other organizers, specially to Cándido Piñeiro, Ramón Rodríguez and Enrique Serrano, who were in charge of almost all necessary organization duties, the Sede Iberoamericana of the University International of Andalusia and its Director Luis C. Contreras, the Local Committee, the many official and private sponsors, listed at the end of this preface, and, of course, the participants themselves without whom there could have been no conference.

Sevilla and Huelva, Spring 2008

The editors

J. M. Delgado and T. Domínguez

SPONSORS

ANDALUSIAN UNIVERSITIES:

University International of Andalusia
University of Almería
University of Cádiz
University of Granada
University of Huelva
University of Jaén
University of Málaga
University Pablo de Olavide
University of Seville

FOUNDATIONS

Fundación Cajasol

OFFICIAL INSTITUTIONS

Ministerio de Educación y Ciencia (Acción Complementaria MTM2006-28259-E)
Consolider Ingenio 2010. Ingenio Mathematica (SARE-C2-0080)
Consejo Social de la Universidad de Huelva
Plan Propio de la Universidad de Sevilla
Vicerrectorado de Investigación de la Universidad de Huelva
Diputación Provincial de Huelva
Ayuntamiento de Palos de la Frontera

ORGANIZING COMMITTEES

SCIENTIFIC/ORGANIZING COMMITTEE

Antonio Aizpuru Tomás	– Universidad de Cádiz
Santiago Díaz Madrigal	– Universidad de Sevilla
Tomás Domínguez Benavides	– Universidad de Sevilla
Daniel Girela Álvarez	– Universidad de Málaga
El Amin Kaidi Lhachmi	– Universidad de Almería
Fernando León Saavedra	– Universidad de Cádiz
Miguel Marano Calzolari	– Universidad de Jaén
Francisco Javier Martín Reyes	– Universidad de Málaga
Juan Francisco Mena Jurado	– Universidad de Granada
Juan Carlos Navarro Pascual	– Universidad de Almería
Rafael Payá Albert	– Universidad de Granada
Carlos Pérez Moreno	– Universidad de Sevilla
Cándido Piñeiro Gómez	– Universidad de Huelva
Francisco Roca Rodríguez	– Universidad de Jaén
Ramón Jaime Rodríguez Álvarez	– Universidad de Huelva
Ángel Rodríguez Palacios	– Universidad de Granada
Luis Rodríguez Piazza	– Universidad de Sevilla
Mª Victoria Velasco Collado	– Universidad de Granada
Antonio Villar Notario	– Universidad Pablo de Olavide

LOCAL ORGANIZING COMMITTEE

Juan Manuel Delgado Sánchez	– Universidad de Huelva
Begoña Marchena González	– Universidad de Huelva
Victoria Martín Márquez	– Universidad de Sevilla
José Antonio Prado Bassas	– Universidad de Sevilla
Enrique Serrano Aguilar	– Universidad de Huelva

CONTENTS

AN INTRODUCTION TO DISCRETE HOLOMORPHIC LOCAL DYNAMICS IN ONE COMPLEX VARIABLE

MARCO ABATE

Dipartimento di Matematica
Università di Pisa
56127 Pisa, Italy
E-mail: abate@dm.unipi.it

1. Introduction

In this survey, by *one–dimensional discrete holomorphic local dynamical system*, we mean a holomorphic function $f: U \to \mathbb{C}$ such that $f(0) = 0$, where $U \subseteq \mathbb{C}$ is an open neighbourhood of 0; we shall also assume that $f \not\equiv \mathrm{id}_U$. We shall denote by End $(\mathbb{C}, 0)$ the set of one–dimensional discrete holomorphic local dynamical systems.

Remark 1.1. Since in this survey we shall only be concerned with the one–dimensional discrete case, we shall often drop the adjectives "one–dimensional" and "discrete" and we shall call an element of End $(\mathbb{C}, 0)$ simply a holomorphic local dynamical system. We shall not discuss at all continuous holomorphic local dynamical systems (e.g., holomorphic ODEs or foliations); however, replacing $\mathbb{C}$ by a complex manifold M and 0 by a point $p \in M$, we recover the general definition of discrete holomorphic local dynamical system in M at p.

Remark 1.2. Since we are mainly concerned with the behaviour of f nearby 0, we shall sometimes replace f by its restriction to some suitable open neighbourhood of 0. It is possible to formalize this fact by using germs of maps and germs of sets at the origin, but for our purposes, it will be enough to use a somewhat less formal approach.

To talk about the dynamics of an $f \in$ End $(\mathbb{C}, 0)$, we need to introduce the iterates of f. If f is defined on the set U then the second iterate $f^2 = f \circ f$ is defined on $U \cap f^{-1}(U)$, which is still an open neighbourhood of the origin. More generally, the k-th iterate $f^k = f \circ f^{k-1}$ is only defined on $U \cap f^{-1}(U) \cap \cdots \cap f^{-(k-1)}(U)$. Thus, it is natural to introduce the *stable set* K_f of f by setting

$$K_f = \bigcap_{k=0}^{\infty} f^{-k}(U).$$

Clearly, $0 \in K_f$ and so, the stable set is never empty (but it can happen that $K_f = \{0\}$; see the next section for an example). The stable set of f is the set of all points $z \in U$ such that the *orbit* $\{f^k(z)\colon k \in \mathbb{N}\}$ is well–defined. If $z \in U \setminus K_f$, we shall say that z (or its orbit) *escapes* from U.

Then the first natural question in local holomorphic dynamics is

Question 1.1. What is the topological structure of K_f?

For instance, when does K_f have non-empty interior? As we shall see in Section 5, holomorphic local dynamical systems such that 0 belongs to the interior of the stable set enjoy special properties.

Remark 1.3. Both the definition of stable set and Question 1.1 (as well as several other definitions or questions we shall meet later on) are topological in character; we might state them for local dynamical systems which are only continuous. As we shall see, however, the *answers* will strongly depend on the holomorphicity of the dynamical system.

Clearly, the stable set K_f is *completely f-invariant*, that is, $f^{-1}(K_f) = K_f$ (this implies, in particular, that $f(K_f) \subseteq K_f$). Therefore, the pair (K_f, f) is a discrete dynamical system in the usual sense and so, the second natural question in local holomorphic dynamics is

Question 1.2. What is the dynamical structure of (K_f, f)?

For instance, what is the asymptotic behaviour of the orbits? Do they converge to the origin, or have they a chaotic behaviour? Is there a dense orbit? Do there exist proper *f-invariant* subsets, that is, sets $L \subset K_f$ such that $f(L) \subseteq L$? If they do exist, what is the dynamics on them?

To answer all these questions, the most efficient way is to replace f by a "dynamically equivalent" but simpler (e.g., linear) map g. In our context, "dynamically equivalent" means "locally conjugated"; and we have at least three kinds of conjugacy to consider.

Let $f_1\colon U_1 \to \mathbb{C}$ and $f_2\colon U_2 \to \mathbb{C}$ be two holomorphic local dynamical system. We shall say that f_1 and f_2 are *holomorphically* (respectively, *topologically*) *locally conjugated* if there are open neighbourhoods $W_1 \subseteq U_1$ and

$W_2 \subseteq U_2$ of the origin and a biholomorphism (respectively, a homeomorphism) $\varphi\colon W_1 \to W_2$ with $\varphi(0) = 0$ such that

$$f_1 = \varphi^{-1} \circ f_2 \circ \varphi \quad \text{on} \quad \varphi^{-1}\left(W_2 \cap f_2^{-1}(W_2)\right) = W_1 \cap f_1^{-1}(W_1).$$

In particular, we have

$$\begin{aligned} f_1^k = \varphi^{-1} \circ f_2^k \circ \varphi \quad \text{on} \qquad & \varphi^{-1}\left(W_2 \cap \cdots \cap f_2^{-(k-1)}(W_2)\right) \\ & = W_1 \cap \cdots \cap f_1^{-(k-1)}(W_1), \end{aligned}$$

for every $k \in \mathbb{N}$ and thus, $K_{f_2|W_2} = \varphi(K_{f_1|W_1})$. So the local dynamics of f_1 is to all purposes equivalent to the local dynamics of f_2.

Whenever we have an equivalence relation in a class of objects, there are classification problems. So the third natural question in local holomorphic dynamics is

Question 1.3. Find a (possibly small) class $\mathcal{F}$ of holomorphic local dynamical systems such that every holomorphic local dynamical system $f \in \operatorname{End}(\mathbb{C}, 0)$ is holomorphically (respectively, topologically) locally conjugated to a (possibly) unique element of $\mathcal{F}$, called the *holomorphic* (respectively, *topological*) *normal form* of f.

Unfortunately, the holomorphic classification is often too complicated to be practical; the family $\mathcal{F}$ of normal forms might be uncountable. A possible replacement is looking for invariants instead of normal forms:

Question 1.4. Find a way to associate a (possibly small) class of (possibly computable) objects, called *invariants*, to any holomorphic local dynamical system f so that two holomorphically conjugated local dynamical systems have the same invariants. The class of invariants is furthermore said *complete* if two holomorphic local dynamical systems are holomorphically conjugated if and only if they have the same invariants.

As remarked before, up to now all the questions we asked make sense for topological local dynamical systems; the next one instead makes sense only for holomorphic local dynamical systems.

A holomorphic local dynamical system is clearly given by an element of $\mathbb{C}_0\{z\}$, the space of converging power series in z without constant terms. The space $\mathbb{C}_0\{z\}$ is a subspace of the space $\mathbb{C}_0[[z]]$ of formal power series without constant terms. An element $\Phi \in \mathbb{C}_0[[z]]$ has an inverse (with respect to composition) still belonging to $\mathbb{C}_0[[z]]$ if and only if its linear part is not zero, that is, if and only if it is not divisible by z^2. We shall then say that two holomorphic local dynamical systems $f_1, f_2 \in \mathbb{C}_0\{z\}$ are *formally conjugated* if there exists an invertible $\Phi \in \mathbb{C}_0[[z]]$ such that $f_1 = \Phi^{-1} \circ f_2 \circ \Phi$ in $\mathbb{C}_0[[z]]$.

It is clear that two holomorphically locally conjugated dynamical systems are both formally and topologically locally conjugated too. On the other hand, we shall see (in Remark 4.2) examples of holomorphic local dynamical systems that are topologically locally conjugated without being neither formally nor holomorphically locally conjugated and (in Remarks 4.2 and 5.3) examples of holomorphic local dynamical systems that are formally conjugated without being neither holomorphically nor topologically locally conjugated. So the last natural question in local holomorphic dynamics we shall deal with is

Question 1.5. Find normal forms and invariants with respect to the relation of formal conjugacy for holomorphic local dynamical systems.

In this survey we shall present some of the main results known on these questions. But before entering the main core of the paper, I would like to thank heartily François Berteloot, Salvatore Coen, Santiago Díaz–Madrigal, Vincent Guedj, Giorgio Patrizio, Mohamad Pouryayevali, Jasmin Raissy, Francesca Tovena and Alekos Vidras, without whom this survey would never has been written.

2. Hyperbolic dynamics

As remarked in the previous section, an one–dimensional discrete holomorphic local dynamical system is given by a converging power series f without constant term:

$$f(z) = a_1 z + a_2 z^2 + a_3 z^3 + \cdots \in \mathbb{C}_0\{z\}.$$

The number $a_1 = f'(0)$ is the *multiplier* of f. Since $a_1 z$ is the best linear approximation of f, it is sensible to expect that the local dynamics of f will be strongly influenced by the value of a_1. We then introduce the following definitions:

- if $|a_1| < 1$ we say that the fixed point 0 is *attracting*;
- if $a_1 = 0$ we say that the fixed point 0 is *superattracting*;
- if $|a_1| > 1$ we say that the fixed point 0 is *repelling*;
- if $|a_1| \neq 0, 1$ we say that the fixed point 0 is *hyperbolic*;
- if $a_1 \in S^1$ is a root of unity we say that the fixed point 0 is *parabolic* (or *rationally indifferent*);

- if $a_1 \in S^1$ is not a root of unity we say that the fixed point 0 is *elliptic* (or *irrationally indifferent*).

Remark 2.1. If $a_1 \neq 0$ then f is locally invertible, that is, there exists $f^{-1} \in \text{End}\,(\mathbb{C}, 0)$ so that $f^{-1} \circ f = f \circ f^{-1} = \text{id}$ where defined. In particular, if 0 is an attracting fixed point for $f \in \text{End}\,(\mathbb{C}, 0)$ with non-zero multiplier then it is a repelling fixed point for the inverse function f^{-1}.

As we shall see in a minute, the dynamics of one–dimensional holomorphic local dynamical systems with a hyperbolic fixed point is pretty elementary; so we start with this case.

Assume first that 0 is attracting (but not superattracting) for the holomorphic local dynamical system $f \in \text{End}\,(\mathbb{C}, 0)$. Then we can write $f(z) = a_1 z + O(z^2)$, with $0 < |a_1| < 1$; hence, we can find a large constant $M > 0$, a small constant $\varepsilon > 0$ and $0 < \delta < 1$ such that if $|z| < \varepsilon$ then

$$|f(z)| \leq (|a_1| + M\varepsilon)|z| \leq \delta |z|. \tag{1}$$

In particular, if Δ_ε is the disk of center 0 and radius ε, we have $f(\Delta_\varepsilon) \subset \Delta_\varepsilon$ for $\varepsilon > 0$ small enough and the stable set of $f|_{\Delta_\varepsilon}$ is Δ_ε itself (in particular, it contains the origin in its interior). Furthermore, since Δ_ε is f-invariant, we can apply (1) to $f(z)$; arguing by induction we get

$$|f^k(z)| \leq \delta^k |z| \to 0 \tag{2}$$

as $k \to +\infty$ and thus, every orbit starting in Δ_ε is attracted by the origin, which is the reason of the name "attracting" for such a fixed point.

If instead 0 is a repelling fixed point, a similar argument (or the observation that 0 is attracting for f^{-1}) shows that for $\varepsilon > 0$ small enough the stable set of $f|_{\Delta_\varepsilon}$ reduces to the origin only: all (non-trivial) orbits escape.

It is also not difficult to find holomorphic and topological normal forms in this case, as shown in the following result, which has marked the beginning of the theory of holomorphic dynamical systems:

Theorem 2.1 (Kœnigs, 1884 [19]). *Let $f \in \text{End}\,(\mathbb{C}, 0)$ be an one–dimensional discrete holomorphic local dynamical system with a hyperbolic fixed point at the origin and let $a_1 \in \mathbb{C}^* \setminus S^1$ be its multiplier. Then:*

(i) f is holomorphically (and hence, formally) locally conjugated to its linear part $g(z) = a_1 z$. The conjugation φ is uniquely determined by the condition $\varphi'(0) = 1$.

(ii) Two such holomorphic local dynamical systems are holomorphically conjugated if and only if they have the same multiplier.

(iii) f is topologically locally conjugated to the map $g_<(z) = z/2$ if $|a_1| < 1$ and to the map $g_>(z) = 2z$ if $|a_1| > 1$.

Proof. Let us assume $0 < |a_1| < 1$; if $|a_1| > 1$, it will suffice to apply the same argument to f^{-1}.

(i) Choose $0 < \delta < 1$ such that $\delta^2 < |a_1| < \delta$. Writing $f(z) = a_1 z + z^2 r(z)$ for a suitable holomorphic germ r, we can find $\varepsilon > 0$ such that $|a_1| + M\varepsilon < \delta$, where $M = \max_{z \in \bar{\Delta}_\varepsilon} |r(z)|$. So we have

$$|f(z) - a_1 z| \leq M|z|^2 \tag{3}$$

that implies (1) and hence, we get (2) for all $z \in \bar{\Delta}_\varepsilon$ and $k \in \mathbb{N}$.

Put $\varphi_k = f^k / a_1^k$; we claim that the sequence $\{\varphi_k\}$ converges to a holomorphic map $\varphi\colon \Delta_\varepsilon \to \mathbb{C}$. Indeed (3) and (2) yield

$$\begin{aligned} |\varphi_{k+1}(z) - \varphi_k(z)| &= \frac{1}{|a_1|^{k+1}} \left| f\left(f^k(z)\right) - a_1 f^k(z) \right| \\ &\leq \frac{M}{|a_1|^{k+1}} |f^k(z)|^2 \leq \frac{M}{|a_1|} \left(\frac{\delta^2}{|a_1|} \right)^k |z|^2 \end{aligned}$$

for all $z \in \bar{\Delta}_\varepsilon$ and so, the telescopic series $\sum_k (\varphi_{k+1} - \varphi_k)$ is uniformly convergent in Δ_ε to $\varphi - \varphi_0$.

Since $\varphi_k'(0) = 1$ for all $k \in \mathbb{N}$, by Weierstrass' theorem we have $\varphi'(0) = 1$ and so, up to possibly shrink ε, we can assume that φ is a biholomorphism with its image. Moreover, we have

$$\varphi\left(f(z)\right) = \lim_{k \to +\infty} \frac{f^k\left(f(z)\right)}{a_1^k} = a_1 \lim_{k \to +\infty} \frac{f^{k+1}(z)}{a_1^{k+1}} = a_1 \varphi(z),$$

that is, $f = \varphi^{-1} \circ g \circ \varphi$, as claimed.

If ψ is another local holomorphic function such that $\psi'(0) = 1$ and $\psi^{-1} \circ g \circ \psi = f$, it follows that $\psi \circ \varphi^{-1}(\lambda z) = \lambda \psi \circ \varphi^{-1}(z)$; comparing the power series expansion of both sides we find $\psi \circ \varphi^{-1} \equiv \mathrm{id}$, that is, $\psi \equiv \varphi$, as claimed.

(ii) Since $f_1 = \varphi^{-1} \circ f_2 \circ \varphi$ implies $f_1'(0) = f_2'(0)$, the multiplier is invariant under holomorphic local conjugation and so, two one–dimensional discrete holomorphic local dynamical systems with a hyperbolic fixed point are holomorphically locally conjugated if and only if they have the same multiplier.

(iii) It suffices to build a topological conjugacy between g and $g_<$ on Δ_ε. First choose a homeomorphism χ between the annulus $\{|a_1|\varepsilon \leq |z| \leq \varepsilon\}$

and the annulus $\{\varepsilon/2 \leq |z| \leq \varepsilon\}$ which is the identity on the outer circle and given by $\chi(z) = z/2a_1$ on the inner circle. Now extend χ by induction to a homeomorphism between the annuli $\{|a_1|^k\varepsilon \leq |z| \leq |a_1|^{k-1}\varepsilon\}$ and $\{\varepsilon/2^k \leq |z| \leq \varepsilon/2^{k-1}\}$ by prescribing

$$\chi(a_1 z) = \frac{1}{2}\chi(z).$$

Putting finally $\chi(0) = 0$, we then get a homeomorphism χ of Δ_ε with itself such that $g = \chi^{-1} \circ g_< \circ \chi$, as required. □

Remark 2.2. The proof of Theorem 2.1(i) relies on a standard trick used to build conjugations in dynamics. Suppose we would like to prove that f and g are conjugated with g invertible. Set $\varphi_k = g^{-k} \circ f^k$, so that

$$\varphi_k \circ f = g \circ \varphi_{k+1}.$$

If the sequence $\{\varphi_k\}$ converges as $k \to +\infty$ to a locally invertible function φ, we automatically have $\varphi \circ f = g \circ \varphi$ and so, φ is the conjugation we were looking for.

Remark 2.3. The proof of Theorem 2.1(iii) uses a standard dynamical trick for building topological conjugations too. Let $f\colon X \to X$ be a continuos closed injective map. A *fundamental domain* for f is a closed set $D \subset X$ with non-empty interior $\mathring{D}$ such that

(i) $X = \bigcup_{k\geq 0} f^k(D)$;
(ii) $f^h(\mathring{D}) \cap f^k(\mathring{D}) = \emptyset$ for all $h \neq k$;
(iii) $f^h(D) \cap f^k(D) \neq \emptyset$ if and only if $|h - k| \leq 1$.

Assume now that you have two continuous closed injective maps $f_1\colon X_1 \to X_1$ and $f_2\colon X_2 \to X_2$ with fundamental domains $D_1 \subset X_1$ and $D_2 \subset X_2$. Assume furthermore that you have a homeomorphism $\chi\colon D_1 \to D_2$ such that

$$\chi\left(f_1(z)\right) = f_2\left(\chi(z)\right) \tag{4}$$

for all $z \in D_1 \cap f_1^{-1}(D_1)$. Then we can extend χ to a homeomorphism between $f_1(D_1)$ and $f_2(D_2)$ by setting $\chi(z) = f_2\left(\chi\left(f_1^{-1}(z)\right)\right)$ for all $z \in f_1(D_1)$; since (4) holds on $D_1 \cap f_1^{-1}(D_1)$, we have obtained a homeomorphism between $D_1 \cup f_1(D_1)$ and $D_2 \cup f(D_2)$ satisfying (4) on $(D_1 \cup f_1(D_1)) \cap f_1^{-1}\left(D_1 \cup f_1(D_1)\right)$. Proceeding in this way, we get a homeomorphism $\chi\colon X_1 \to X_2$ satisfying $\chi \circ f_1 = f_2 \circ \chi$, as desired.

Remark 2.4. Notice that $g_<(z) = \frac{1}{2}z$ and $g_>(z) = 2z$ cannot be topologically conjugated, because (for instance) the stable set of $g_<$ is open whereas the stable set of $g_>$ only contains the origin.

3. Superattracting dynamics

Let us now study the superattracting case. If 0 is a superattracting point for an $f \in \mathrm{End}\,(\mathbb{C},0)$, we can write

$$f(z) = a_r z^r + a_{r+1} z^{r+1} + \cdots$$

with $a_r \neq 0$; the number $r \geq 2$ is the *order* of the superattracting point. An argument similar to the one described in the previous section shows that for $\varepsilon > 0$ small enough, the stable set of $f|_{\Delta_\varepsilon}$ is still all of Δ_ε and the orbits converge (faster than in the attracting case) to the origin. Furthermore, we can prove the following

Theorem 3.1 (Böttcher, 1904 [2]). *Let $f \in \mathrm{End}\,(\mathbb{C},0)$ be an one-dimensional holomorphic local dynamical system with a superattracting fixed point at the origin and let $r \geq 2$ be its order. Then:*

(i) f is holomorphically (and hence formally and topologically) locally conjugated to the map $g(z) = z^r$.
(ii) two such holomorphic local dynamical systems are holomorphically (or topologically or formally) conjugated if and only if they have the same order.

Proof. First of all, up to a linear conjugation $z \mapsto \mu z$ with $\mu^{r-1} = a_r$, we can assume $a_r = 1$.

Now write $f(z) = z^r h_1(z)$ for some holomorphic germ h_1 with $h_1(0) = 1$. By induction, it is easy to see that we can write $f^k(z) = z^{r^k} h_k(z)$ for a suitable holomorphic germ h_k with $h_k(0) = 1$. Furthermore, the equalities $f \circ f^{k-1} = f^k = f^{k-1} \circ f$ yield

$$h_{k-1}(z)^r h_1\left(f^{k-1}(z)\right) = h_k(z) = h_1(z)^{r^{k-1}} h_{k-1}\left(f(z)\right). \tag{5}$$

Choose $0 < \delta < 1$. Then we can clearly find $1 > \varepsilon > 0$ such that $M\varepsilon < \delta$, where $M = \max_{z \in \bar{\Delta}_\varepsilon} |h_1(z)|$; we can also assume that $h_1(z) \neq 0$ for all $z \in \bar{\Delta}_\varepsilon$. Since

$$|f(z)| \leq M|z|^r < \delta|z|^{r-1} \quad \forall z \in \bar{\Delta}_\varepsilon,$$

we have $f(\Delta_\varepsilon) \subset \Delta_\varepsilon$, as anticipated before.

We also remark that (5) implies that each h_k is well–defined and never vanishing on $\bar{\Delta}_\varepsilon$. So, for every $k \geq 1$, we can choose a unique ψ_k holomorphic in Δ_ε such that $\psi_k(z)^{r^k} = h_k(z)$ on Δ_ε and with $\psi_k(0) = 1$.

Set $\varphi_k(z) = z\psi_k(z)$ so that $\varphi_k'(0) = 1$ and $\varphi_k(z)^{r^k} = f_k(z)$ on Δ_ε. We claim that the sequence $\{\varphi_k\}$ converges to a holomorphic function φ on Δ_ε. Indeed, we have

$$\left|\frac{\varphi_{k+1}(z)}{\varphi_k(z)}\right| = \left|\frac{\psi_{k+1}(z)^{r^{k+1}}}{\psi_k(z)^{r^{k+1}}}\right|^{1/r^{k+1}} = \left|\frac{h_{k+1}(z)}{h_k(z)^r}\right|^{1/r^{k+1}} = \left|h_1\left(f^k(z)\right)\right|^{1/r^{k+1}}$$

$$= \left|1 + O\left(|f^k(z)|\right)\right|^{1/r^{k+1}} = 1 + \frac{1}{r^{k+1}} O\left(|f^k(z)|\right) = 1 + O\left(\frac{1}{r^{k+1}}\right),$$

and so, the telescopic product $\prod_k(\varphi_{k+1}/\varphi_k)$ converges to φ/φ_1 uniformly in Δ_ε.

Since $\varphi_k'(0) = 1$ for all $k \in \mathbb{N}$, we have $\varphi'(0) = 1$ and so, up to possibly shrink ε, we can assume that φ is a biholomorphism with its image. Moreover, we have

$$\begin{aligned}\varphi_k\left(f(z)\right)^{r^k} &= f(z)^{r^k}\psi_k\left(f(z)\right)^{r^k} = z^{r^{k+1}} h_1(z)^{r^k} h_k\left(f(z)\right) \\ &= z^{r^{k+1}} h_{k+1}(z) = \left[\varphi_{k+1}(z)^r\right]^{r^k},\end{aligned}$$

and thus, $\varphi_k \circ f = [\varphi_{k+1}]^r$. Passing to the limit, we get $f = \varphi^{-1} \circ g \circ \varphi$, as claimed.

Finally, (ii) follows because z^r and z^s are locally topologically (or formally) conjugated if and only if $r = s$. □

Therefore, the one–dimensional local dynamics about a hyperbolic or superattracting fixed point is completely clear; let us now discuss what happens about a parabolic fixed point.

4. Parabolic dynamics

Let $f \in \mathrm{End}\,(\mathbb{C}, 0)$ be a (non-linear) holomorphic local dynamical system with a parabolic fixed point at the origin. Then we can write

$$f(z) = e^{2i\pi p/q} z + a_{r+1} z^{r+1} + a_{r+2} z^{r+2} + \cdots, \tag{6}$$

with $a_{r+1} \neq 0$, where $p/q \in \mathbb{Q} \cap [0, 1)$ is the *rotation number* of f and the number $r + 1 \geq 2$ is the *multiplicity* of f at the fixed point.

The first observation is that such a dynamical system is never locally conjugated to its linear part, not even topologically, unless it is of finite order:

Proposition 4.1. *Let $f \in \mathrm{End}\,(\mathbb{C},0)$ be a holomorphic local dynamical system with multiplier λ and assume that λ is a primitive root of the unity of order q. Then f is holomorphically (or topologically or formally) linearizable if and only if $f^q \equiv \mathrm{id}$.*

Proof. Indeed, if $\varphi^{-1} \circ f \circ \varphi(z) = e^{2\pi i p/q} z$ we get $\varphi^{-1} \circ f^q \circ \varphi \equiv \mathrm{id}$, that is, $f^q \equiv \mathrm{id}$. Conversely, assume that $f^q \equiv \mathrm{id}$ and set

$$\varphi(z) = \frac{1}{q} \sum_{j=0}^{q-1} \frac{f^j(z)}{\lambda^j}.$$

Then it is easy to check that $\varphi'(0) = 1$ and $\varphi \circ f(z) = \lambda \varphi(z)$ and so, f is holomorphically (and topologically and formally) linearizable. □

In particular, if the rotation number is 0 (that is, the multiplier is 1 and we shall say that f is *tangent to the identity*), then f *cannot* be locally conjugated to the identity (unless it was the identity to begin with, which is not a very interesting case dynamically speaking). More precisely, the stable set of such an f is never a neighbourhood of the origin. To understand why, let us first consider a map of the form

$$f(z) = z(1 + az^r)$$

for some $a \neq 0$. Let $v \in S^1 \subset \mathbb{C}$ be such that av^r is real and positive. Then, for any $c > 0$, we have

$$f(cv) = c(1 + c^r a v^r) v \in \mathbb{R}^+ v;$$

moreover, $|f(cv)| > |cv|$. In other words, the half–line $\mathbb{R}^+ v$ is f-invariant and repelled from the origin, that is, $K_f \cap \mathbb{R}^+ v = \emptyset$. Conversely, if av^r is real and negative then it is easy to see that the segment $[0, |a|^{-1/r}]v$ is f-invariant and attracted by the origin. So K_f neither is a neighbourhood of the origin nor reduces to $\{0\}$.

This example suggests the following definition. Let $f \in \mathrm{End}\,(\mathbb{C},0)$ be of the form

$$f(z) = z + a_{r+1} z^{r+1} + a_{r+2} z^{r+2} + \cdots . \tag{7}$$

Then a unit vector $v \in S^1$ is an *attracting* (respectively, *repelling*) *direction* for f at the origin if $a_{r+1} v^r$ is real and negative (respectively, positive). Clearly, there are r equally spaced attracting directions, separated by r equally spaced repelling directions: if $a_{r+1} = |a_{r+1}| e^{i\alpha}$ then $v = e^{i\theta}$ is attracting (respectively, repelling) if and only if

$$\theta = \frac{2k+1}{r}\pi - \frac{\alpha}{r} \qquad \left(\text{respectively, } \theta = \frac{2k}{r}\pi - \frac{\alpha}{r}\right).$$

Furthermore, a repelling (attracting) direction for f is attracting (repelling) for f^{-1}, which is defined in a neighbourhood of the origin.

It turns out that to every attracting direction is associated a connected component of $K_f \setminus \{0\}$. Let $v \in S^1$ be an attracting direction for an f tangent to the identity. The *basin* centered at v is the set of points $z \in K_f \setminus \{0\}$ such that $f^k(z) \to 0$ and $f^k(z)/|f^k(z)| \to v$ (notice that, up to shrinking the domain of f, we can assume that $f(z) \neq 0$ for all $z \in K_f \setminus \{0\}$). If z belongs to the basin centered at v, we shall say that the orbit of z *tends to* 0 *tangent to* v.

A slightly more specialized (but more useful) object is the following: an *attracting petal* centered at an attracting direction v is an open simply connected f-invariant set $P \subseteq K_f \setminus \{0\}$ such that a point $z \in K_f \setminus \{0\}$ belongs to the basin centered at v if and only if its orbit intersects P. In other words, the orbit of a point tends to 0 tangent to v if and only if it is eventually contained in P. A *repelling petal* (centered at a repelling direction) is an attracting petal for the inverse of f.

The basins centered at the attracting directions are exactly the connected components of $K_f \setminus \{0\}$, as shown in the *Leau–Fatou flower theorem*:

Theorem 4.1 (Leau, 1897 [20]; Fatou, 1919-20 [12–14]). *Let* $f \in \operatorname{End}(\mathbb{C}, 0)$ *be a holomorphic local dynamical system tangent to the identity with multiplicity* $r + 1 \geq 2$ *at* 0. *Let* $v_1, v_3, \ldots, v_{2r-1} \in S^1$ *be the* r *attracting directions of* f *at the origin and* $v_2, v_4, \ldots, v_{2r} \in S^1$ *the* r *repelling directions. Then*

- *(i) There exists for each attracting (repelling) direction* v_{2j-1} (v_{2j}) *an attracting (repelling) petal* P_{2j-1} (P_{2j}) *so that the union of these* $2r$ *petals together with the origin forms a neighbourhood of the origin. Furthermore, the* $2r$ *petals are arranged cyclically so that two petals intersect if and only if the angle between their central directions is* π/r.
- *(ii)* $K_f \setminus \{0\}$ *is the (disjoint) union of the basins centered at the* r *attracting directions.*
- *(iii) If* B *is a basin centered at an attracting direction, there exists a function* $\varphi\colon B \to \mathbb{C}$ *such that* $\varphi \circ f(z) = \varphi(z) + 1$ *for all* $z \in B$. *Furthermore, if* P *is the petal constructed in part (i) then* $\varphi|_P$ *is a biholomorphism with an open subset of the complex plane containing a right half-plane — and so,* $f|_P$ *is holomorphically conjugated to the translation* $z \mapsto z + 1$.

Proof. Up to a linear conjugation, we can assume that $a_{r+1} = -1$, so that the attracting directions are the r-th roots of unity. For any $\delta > 0$, the set $\{z \in \mathbb{C}\colon |z^r - \delta| < \delta\}$ has exactly r connected components, each one symmetric with respect to a different r-th root of unity; it will turns out that, for δ small enough, these connected components are attracting petals of f, even though to get a pointed neighbourhood of the origin we shall need larger petals.

For $j = 1, 3, \ldots, 2r-1$, let $\Sigma_j \subset \mathbb{C}^*$ denote the sector centered about the attractive direction v_j and bounded by two consecutive repelling directions, that is,

$$\Sigma_j = \left\{ z \in \mathbb{C}^* : \frac{2j-3}{r}\pi < \arg z < \frac{2j-1}{r}\pi \right\}.$$

Notice that each Σ_j contains a unique connected component $P_{j,\delta}$ of the set $\{z \in \mathbb{C}\colon |z^r - \delta| < \delta\}$; moreover, $P_{j,\delta}$ is tangent at the origin to the sector centered about v_j of amplitude π/r.

The main technical trick in this proof consists in transferring the setting to a neighbourhood of infinity in the Riemann sphere $\mathbb{P}^1(\mathbb{C})$. For $j = 1, 3, \ldots, 2r-1$, the function $\psi\colon \mathbb{C}^* \to \mathbb{C}^*$ given by

$$\psi(z) = \frac{1}{rz^r}$$

is a biholomorphism between Σ_j and $\mathbb{C}^* \setminus \mathbb{R}^-$ with inverse of the form $\psi^{-1}(w) = (rw)^{-1/r}$, suitably choosing the r-th root. Furthermore, $\psi(P_{j,\delta})$ is the right half–plane $H_\delta = \{w \in \mathbb{C}\colon \operatorname{Re} w > 1/(2r\delta)\}$.

When $|w|$ is so large that $\psi^{-1}(w)$ belongs to the domain of definition of f, the composition $F = \psi \circ f \circ \psi^{-1}$ makes sense and we have

$$F(w) = w + 1 + O(w^{-1/r}). \tag{8}$$

Thus, to study the dynamics of f in a neighbourhood of the origin in Σ_j, it suffices to study the dynamics of F in a neighbourhood of infinity.

The first observation is that if $\operatorname{Re} w$ is large enough then

$$\operatorname{Re} F(w) > \operatorname{Re} w + \frac{1}{2};$$

this implies that, for δ small enough, H_δ is F-invariant (and thus, $P_{j,\delta}$ is f-invariant). Furthermore, by induction one has

$$\operatorname{Re} F^k(w) > \operatorname{Re} w + \frac{k}{2} \quad \forall w \in H_\delta \tag{9}$$

which implies that $F^k(w) \to \infty$ in H_δ (and $f^k(z) \to 0$ in $P_{j,\delta}$) as $k \to \infty$.

Now we claim that the argument of $w_k = F^k(w)$ tends to zero. Indeed, (8) yields

$$\frac{w_k}{k} = \frac{w}{k} + 1 + \frac{1}{k}\sum_{l=0}^{k-1} O(w_l^{-1/r});$$

so Cesaro's theorem on the averages of a converging sequence implies

$$\frac{w_k}{k} \to 1, \tag{10}$$

and thus, $\arg w_k \to 0$ as $k \to \infty$. Going back to $P_{j,\delta}$, this implies that $f^k(z)/|f^k(z)| \to v_j$ for every $z \in P_{j,\delta}$. Since furthermore $P_{j,\delta}$ is centered about v_j, every orbit converging to 0 tangent to v_j must intersect $P_{j,\delta}$ and thus, we have proved that $P_{j,\delta}$ is an attracting petal.

Arguing in the same way with f^{-1} we get repelling petals; unfortunately, these petals are too small to obtain a full pointed neighbourhood of the origin. In fact, as remarked before, each $P_{j,\delta}$ is contained in a sector centered about v_j of amplitude π/r; therefore, the repelling and attracting petals obtained in this way do not intersect but are tangent to each other. We need larger petals.

So our aim is to find an f-invariant subset $\widetilde{P}_j$ of Σ_j containing $P_{j,\delta}$ and which is tangent at the origin to a sector centered about v_j of amplitude strictly greater than π/r. To do so, first of all remark that there are $R, C > 0$ such that

$$|F(w) - w - 1| \le \frac{C}{|w|^{1/r}} \tag{11}$$

as soon as $|w| > R$. Choose $\varepsilon \in (0,1)$ and select $\delta > 0$ so that $|w| > 1/(2r\delta)$ implies

$$|F(w) - w - 1| \le \varepsilon/2.$$

Set $M_\varepsilon = \sqrt{1+\varepsilon^2}/(2r\delta)$ and let

$$\widetilde{H}_\varepsilon = \{w \in \mathbb{C} \colon \varepsilon|\mathrm{Im}\, w| > -\mathrm{Re}\, w + M_\varepsilon\} \cup H_\delta;$$

in particular, $|w| > 1/(2r\delta)$ for all $w \in \widetilde{H}_\varepsilon$. If $w \in \widetilde{H}_\varepsilon$ we have

$$\mathrm{Re}\, F(w) > \mathrm{Re}\, w + 1 - \varepsilon/2 \qquad \text{and} \qquad |\mathrm{Im}\, F(w) - \mathrm{Im}\, w| < \varepsilon/2; \tag{12}$$

it is then easy to check that $F(\widetilde{H}_\varepsilon) \subset \widetilde{H}_\varepsilon$ and that every orbit starting in $\widetilde{H}_\varepsilon$ must eventually enter H_δ. Therefore, $\widetilde{P}_j = \psi^{-1}(\widetilde{H}_\varepsilon)$ is as required and we have proved (i).

To prove (ii), we need a further property of $\widetilde{H}_\varepsilon$. Since

$$f^{-1}(z) = z + z^{r+1} + O(z^{r+2}),$$

we have

$$F^{-1}(w) = w - 1 + O(w^{-1/r});$$

up to decreasing δ, we can thus assume that $|F^{-1}(w) - w + 1| < \varepsilon/2$ on $\widetilde{H}_\varepsilon$. But then, if $w \in \widetilde{H}_\varepsilon$ we have

$$\operatorname{Re} F^{-1}(w) < \operatorname{Re} w - 1 + \frac{\varepsilon}{2}$$

and

$$\varepsilon|\operatorname{Im} F^{-1}(w)| + \operatorname{Re} F^{-1}(w) < \varepsilon|\operatorname{Im} w| + \operatorname{Re} w - \left(1 - \frac{\varepsilon(1+\varepsilon)}{2}\right);$$

this means that every inverse orbit *must* eventually leave $\widetilde{H}_\varepsilon$.

Coming back to the z-plane, we have thus proved that every (forward) orbit of f must eventually leave any repelling petal. So if $z \in K_f \setminus \{0\}$, where the stable set is computed working in the neighbourhood of the origin constructed in part (i), the orbit of z must eventually land in an attracting petal and thus, z belongs to a basin centered at one of the r attracting directions — and (ii) is proved.

To prove (iii), first of all notice that

$$|F'(w) - 1| \leq \frac{2^{1+1/r}C}{|w|^{1+1/r}} \tag{13}$$

in $\widetilde{H}_\varepsilon$. Indeed, (11) says that if $|w| > 1/(2r\delta)$ then the function $w \mapsto F(w) - w - 1$ sends the disk of center w and radius $|w|/2$ into the disk of center the origin and radius $C/(|w|/2)^{1/r}$ for some $C > 0$; inequality (13) then follows from the Cauchy estimates on the derivative.

Now choose $w_0 \in H_\delta$ and set $\tilde{\varphi}_k(w) = F^k(w) - F^k(w_0)$. Given $w \in \widetilde{H}_\varepsilon$, as soon as $k \in \mathbb{N}$ is so large that $F^k(w) \in H_\delta$, we can apply Lagrange's theorem to the segment from $F^k(w_0)$ to $F^k(w)$ to get a $t_k \in [0,1]$ such that

$$\begin{aligned}
\left|\frac{\tilde{\varphi}_{k+1}(w)}{\tilde{\varphi}_k(w)} - 1\right| &= \left|\frac{F\left(F^k(w)\right) - F^k\left(F^k(w_0)\right)}{F^k(w) - F^k(w_0)} - 1\right| \\
&= \left|F'\left(t_k F^k(w) + (1-t_k)F^k(w_0)\right) - 1\right| \\
&\leq \frac{2^{1+1/r}C}{\min\{\operatorname{Re}|F^k(w), \operatorname{Re}|F^k(w_0)|\}^{1+1/r}} \leq \frac{C'}{k^{1+1/r}},
\end{aligned}$$

where we used (13) and (12) and the constant C' is uniform on compact subsets of $\widetilde{H}_\varepsilon$ (and it can be chosen uniform on H_δ).

As a consequence, the telescopic product $\prod_k \tilde{\varphi}_{k+1}/\tilde{\varphi}_k$ converges uniformly on compact subsets of $\widetilde{H}_\varepsilon$ (and uniformly on H_δ) and thus, the sequence $\tilde{\varphi}_k$ converges uniformly on compact subsets to a holomorphic function $\tilde{\varphi}\colon \widetilde{H}_\varepsilon \to \mathbb{C}$. Since we have

$$\begin{aligned}\tilde{\varphi}_k \circ F(w) &= F^{k+1}(w) - F^k(w_0) = \tilde{\varphi}_{k+1}(w) + F\left(F^k(w_0)\right) - F^k(w_0)\\ &= \tilde{\varphi}_{k+1}(w) + 1 + O\left(|F^k(w_0)|^{-1/r}\right),\end{aligned}$$

it follows that

$$\tilde{\varphi} \circ F(w) = \tilde{\varphi}(w) + 1$$

on $\widetilde{H}_\varepsilon$. In particular, $\tilde{\varphi}$ is not constant; being the limit of injective functions, by Hurwitz's theorem, it is injective.

We now prove that the image of $\tilde{\varphi}$ contains a right half–plane. First of all, we claim that

$$\lim_{\substack{|w|\to+\infty \\ w\in H_\delta}} \frac{\tilde{\varphi}(w)}{w} = 1. \tag{14}$$

Indeed, choose $\eta > 0$. Since the convergence of the telescopic product is uniform on H_δ, we can find $k_0 \in \mathbb{N}$ such that

$$\left|\frac{\tilde{\varphi}(w) - \tilde{\varphi}_{k_0}(w)}{w - w_0}\right| < \frac{\eta}{2}$$

on H_δ. Furthermore, we have

$$\left|\frac{\tilde{\varphi}_{k_0}(w)}{w - w_0} - 1\right| = \left|\frac{k_0 + \sum_{j=0}^{k_0-1} O(|F^j(w)|^{-1/r}) + w_0 - F^{k_0}(w_0)}{w - w_0}\right| = O(|w|^{-1})$$

on H_δ; therefore, we can find $R > 0$ such that

$$\left|\frac{\tilde{\varphi}(w)}{w - w_0} - 1\right| < \eta$$

as soon as $|w| > R$ in H_δ.

Equality (14) clearly implies that $(\tilde{\varphi}(w) - w^o)/(w - w^o) \to 1$ as $|w| \to +\infty$ in H_δ for any $w^o \in \mathbb{C}$. But this means that if $\operatorname{Re} w^o$ is large enough then the difference between the variation of the argument of $\tilde{\varphi} - w^o$ along a suitably small closed circle around w^o and the variation of the argument of $w - w^o$ along the same circle will be less than 2π — and thus, it will be zero. Then the principle of the argument implies that $\tilde{\varphi} - w^o$ and $w - w^o$ have the same number of zeroes inside that circle and thus, $w^o \in \tilde{\varphi}(H_\delta)$, as required.

So setting $\varphi = \tilde{\varphi} \circ \psi$, we have defined a function φ with the required properties on $\widetilde{P}_j$. To extend it to the whole basin B, it suffices to put

$$\varphi(z) = \varphi\left(f^k(z)\right) - k,$$

where $k \in \mathbb{N}$ is the first integer such that $f^k(z) \in \widetilde{P}_j$. □

Remark 4.1. It is possible to construct petals that cannot be contained in any sector strictly smaller than Σ_j. To do so, we need an F-invariant subset $\widehat{H}_\varepsilon$ of $\mathbb{C}^* \setminus \mathbb{R}^-$ containing $\widetilde{H}_\varepsilon$ and containing eventually every half–line issuing from the origin (but $\mathbb{R}^-$). For $M >> 1$ and $C > 0$ large enough, replace the straight lines bounding $\widetilde{H}_\varepsilon$ on the left of Re $w = M_\varepsilon$ by the curves

$$|\mathrm{Im}\, w| = \begin{cases} C \log |\mathrm{Re}\, w| & \text{if } r = 1, \\ C|\mathrm{Re}\, w|^{1-1/r} & \text{if } r > 1. \end{cases}$$

Then it is not too difficult to check that the domain $\widehat{H}_\varepsilon$ so obtained is as desired (see [7]).

So we have a complete description of the dynamics in the neighbourhood of the origin. Actually, Camacho has pushed this argument even further, obtaining a complete topological classification of one–dimensional discrete holomorphic local dynamical systems tangent to the identity:

Theorem 4.2 (Camacho, 1978 [6]; Shcherbakov, 1982 [34]). *Let $f \in$ End $(\mathbb{C}, 0)$ be a holomorphic local dynamical system tangent to the identity with multiplicity $r+1$ at the fixed point. Then f is topologically locally conjugated to the map*

$$z \mapsto z - z^{r+1}.$$

The formal classification is simple too, though different and it can be obtained with an easy computation:

Proposition 4.2. *Let $f \in$ End $(\mathbb{C}, 0)$ be a holomorphic local dynamical system tangent to the identity with multiplicity $r+1$ at the fixed point. Then f is formally conjugated to the map*

$$g(z) = z - z^{r+1} + \beta z^{2r+1}, \tag{15}$$

where β is a formal (and holomorphic) invariant given by

$$\beta = \frac{1}{2\pi i} \int_\gamma \frac{dz}{z - f(z)}, \tag{16}$$

where the integral is taken over a small positive loop γ about the origin.

Proof. A computation shows that if f is given by (15) then β is given by the integral (16). Conversely, let φ be a local biholomorphism fixing the origin and set $F = \varphi^{-1} \circ f \circ \varphi$. Then

$$\frac{1}{2\pi i} \int_\gamma \frac{dz}{z - f(z)} = \frac{1}{2\pi i} \int_{\varphi^{-1} \circ \gamma} \frac{\varphi'(w)\, dw}{\varphi(w) - f(\varphi(w))}$$
$$= \frac{1}{2\pi i} \int_{\varphi^{-1} \circ \gamma} \frac{\varphi'(w)\, dw}{\varphi(w) - \varphi(F(w))}.$$

Now, we can clearly find M, $M_1 > 0$ such that

$$\left| \frac{1}{w - F(w)} - \frac{\varphi'(w)}{\varphi(w) - \varphi(F(w))} \right|$$
$$= \frac{1}{|\varphi(w) - \varphi(F(w))|} \left| \frac{\varphi(w) - \varphi(F(w))}{w - F(w)} - \varphi'(w) \right|$$
$$\leq M \frac{|w - F(w)|}{|\varphi(w) - \varphi(F(w))|} \leq M_1$$

in a neighbourhood of the origin, where the last inequality follows from the fact that $\varphi'(0) \neq 0$. This means that the two meromorphic functions $1/(w - F(w))$ and $\varphi'(w)/(\varphi(w) - \varphi((F(w)))$ differ by a holomorphic function; so they have the same integral along any small loop surrounding the origin and

$$\frac{1}{2\pi i} \int_\gamma \frac{dz}{z - f(z)} = \frac{1}{2\pi i} \int_{\varphi^{-1} \circ \gamma} \frac{dw}{w - F(w)},$$

as claimed.

To prove that f is formally conjugated to g, let us first take a local formal change of coordinates φ of the form

$$\varphi(z) = z + \mu z^d + O_{d+1} \tag{17}$$

with $\mu \neq 0$ and where we are writing O_{d+1} instead of $O(z^{d+1})$. It follows that $\varphi^{-1}(z) = z - \mu z^d + O_{d+1}$, $(\varphi^{-1})'(z) = 1 - d\mu z^{d-1} + O_d$ and $(\varphi^{-1})^{(j)} =$

O_{d-j} for all $j \geq 2$. Then, using the Taylor expansion of φ^{-1}, we get

$$\varphi^{-1} \circ f \circ \varphi(z) = \varphi^{-1}\left(\varphi(z) + \sum_{j \geq r+1} a_j \varphi(z)^j\right)$$

$$= z + (\varphi^{-1})'(\varphi(z)) \sum_{j \geq r+1} a_j z^j (1 + \mu z^{d-1} + O_d)^j + O_{d+2r}$$

$$= z + [1 - d\mu z^{d-1} + O_d] \sum_{j \geq r+1} a_j z^j (1 + j\mu z^{d-1} + O_d) + O_{d+2r}$$

$$= z + a_{r+1} z^{r+1} + \cdots + a_{r+d-1} z^{r+d-1}$$

$$+[a_{r+d} + (r+1-d)\mu a_{r+1}] z^{r+d} + O_{r+d+1}. \tag{18}$$

This means that if $d \neq r+1$ we can use a polynomial change of coordinates of the form $\varphi(z) = z + \mu z^d$ to remove the term of degree $r + d$ from the Taylor expansion of f without changing lower degree terms.

So, to conjugate f to g, it suffices to use a linear change of coordinates to get $a_{r+1} = -1$ and then apply a sequence of change of coordinates of the form $\varphi(z) = z + \mu z^d$ to kill all the terms in the Taylor expansion of f but the term of degree z^{2r+1}.

Finally, formula (18) also shows that two maps of the form (15) with different β cannot be formally conjugated and we are done. □

The number β given by (16) is called *index* of f at the fixed point.

The holomorphic classification is much more complicated: as shown by Voronin ([36]) and Écalle ([10–11]) in 1981, it depends on functional invariants. We shall now (very) roughly describe it; see [17,18,21,22] (and the original papers) for details.

Let $f \in \mathrm{End}\,(\mathbb{C}, 0)$ be tangent to the identity with multiplicity $r + 1$ at the fixed point; up to a linear change of coordinates, we can assume that $a_{r+1} = 1$. Let $P_1, \ldots, P_{2r}$ be a set of petals as in Theorem 4.1(i) chosen so that P_{2r} is centered on the positive real semiaxis and the others are arranged cyclically counterclockwise. Denote by H_j the biholomorphism conjugating $f|_{P_j}$ to the shift $z \mapsto z + 1$ in either a right (if j is odd) or left (if j is even) half–plane given by Theorem 4.1(iii) — applied to f^{-1} for the repelling petals. If we moreover require that

$$H_j(z) = -\frac{1}{rz^r} + \beta \log z + o(1), \tag{19}$$

where β is the index of f at the origin then H_j is uniquely determined. Thus, in the sets $H_j(P_j \cap P_{j+1})$ we can consider the composition $\widetilde{\Phi}_j = H_{j+1} \circ H_j^{-1}$. It is easy to check that $\widetilde{\Phi}_j(w+1) = \widetilde{\Phi}_j(w) + 1$ for $j = 1, \ldots, 2r-1$, and thus, $\psi_j = \widetilde{\Phi}_j - \mathrm{id}$ is a 1-periodic holomorphic function (for $j = 2r$, we need to take $\psi_{2r} = \widetilde{\Phi}_{2r} - \mathrm{id} + 2\pi i \beta$ to get a 1-periodic function). Hence, each ψ_j can be extended to a suitable upper (if j is odd) or lower (if j is even) half–plane. Furthermore, it is possible to prove that the functions $\psi_1, \ldots, \psi_{2r}$ are exponentially decreasing, that is, they are bounded by $\exp(-c|w|)$ as $|\mathrm{Im}\, w| \to +\infty$ for a suitable $c > 0$ depending on f.

Now, if we replace f by a holomorphic local conjugate $g = h^{-1} \circ f \circ h$ and denote by G_j the corresponding biholomorphisms, it turns out that

$$H_j \circ G_j^{-1} = \mathrm{id} + a$$

for a suitable $a \in \mathbb{C}$ independent of j. This suggests the introduction of an equivalence relation on the set of $2r$-uple of functions of the kind $(\psi_1, \ldots, \psi_{2r})$.

Let M_r denote the set of $2r$-uple of holomorphic 1-periodic functions $\psi = (\psi_1, \ldots, \psi_{2r})$ with ψ_j defined in a suitable upper (if j is odd) or lower (if j is even) half–plane and exponentially decreasing when $|\mathrm{Im}\, w| \to +\infty$. We shall say that $\psi, \tilde{\psi} \in M_r$ are *equivalent* if there is $a \in \mathbb{C}$ such that $\tilde{\psi}_j = \psi_j \circ (\mathrm{id} + a)$ for $j = 1, \ldots, 2r$. We denote by $\mathcal{M}_r$ the set of all equivalence classes.

The procedure described above allows us to associate to any $f \in \mathrm{End}\,(\mathbb{C}, 0)$ tangent to the identity with multiplicity $r+1$ at the fixed point an element $\mu_f \in \mathcal{M}_r$ called the *sectorial invariant*. Then the holomorphic classification proved by Écalle and Voronin is

Theorem 4.3 (Écalle, 1981 [10,11]; Voronin, 1981 [36]). *Let f, $g \in \mathrm{End}\,(\mathbb{C}, 0)$ be two holomorphic local dynamical systems tangent to the identity. Then f and g are holomorphically locally conjugated if and only if they have the same multiplicity, the same index and the same sectorial invariant. Furthermore, for any $r \geq 1$, $\beta \in \mathbb{C}$ and $\mu \in \mathcal{M}_r$ there exists $f \in \mathrm{End}\,(\mathbb{C}, 0)$ tangent to the identity with multiplicity $r+1$, index β and sectorial invariant μ.*

Remark 4.2. In particular, holomorphic local dynamical systems tangent to the identity give examples of local dynamical systems that are topologically conjugated without being neither holomorphically nor formally conjugated and of local dynamical systems that are formally conjugated without being holomorphically conjugated.

Finally, if $f \in \text{End}(\mathbb{C}, 0)$ satisfies $a_1 = e^{2\pi i p/q}$ then f^q is tangent to the identity. Therefore, we can apply the previous results to f^q and then infer informations about the dynamics of the original f. We list here a few results; see [6,10,11,23,24,36] for proofs and further details.

Proposition 4.3. *Let $f \in \text{End}(\mathbb{C}, 0)$ be a holomorphic local dynamical system with multiplier λ and assume that λ is a primitive root of the unity of order q. Assume that $f^q \not\equiv \text{id}$. Then there exist $n \geq 1$ and $c \in \mathbb{C}$ such that f is formally conjugated to*

$$g(z) = \lambda z + z^{nq+1} + cz^{2nq+1}.$$

Theorem 4.4 (Camacho). *Let $f \in \text{End}(\mathbb{C}, 0)$ be a holomorphic local dynamical system with multiplier λ and assume that λ is a primitive root of the unity of order q. Assume that $f^q \not\equiv \text{id}$. Then there exist $n \geq 1$ such that f is topologically conjugated to*

$$g(z) = \lambda z + z^{nq+1}.$$

Theorem 4.5 (Leau–Fatou). *Let $f \in \text{End}(\mathbb{C}, 0)$ be a holomorphic local dynamical system with multiplier λ and assume that λ is a primitive root of the unity of order q. Assume that $f^q \not\equiv \text{id}$. Then there exist $n \geq 1$ such that f^q has multiplicity $nq+1$ and f acts on the attracting (respectively, repelling) petals of f^q as a permutation composed by n disjoint cycles. Finally, $K_f = K_{f^q}$.*

5. Elliptic dynamics

We are left with the elliptic case:

$$f(z) = e^{2\pi i\theta} z + a_2 z^2 + \cdots \in \mathbb{C}_0\{z\} \tag{20}$$

with $\theta \notin \mathbb{Q}$. It turns out that the local dynamics depends mostly on the numerical properties of θ. More precisely, for a full measure subset B of $\theta \in [0,1] \setminus \mathbb{Q}$ all holomorphic local dynamical systems of the form (20) are *holomorphically linearizable*, that is, holomorphically locally conjugated to their (common) linear part, the irrational rotation $z \mapsto e^{2\pi i\theta} z$. Conversely, the complement $[0,1] \setminus B$ is a G_δ-dense set and for all $\theta \in [0,1] \setminus B$ the quadratic polynomial $z \mapsto z^2 + e^{2\pi i\theta} z$ is not holomorphically linearizable. This is the gist of the results due to Cremer, Siegel, Bryuno and Yoccoz we are going to describe in this section.

The first worthwhile observation in this setting is that it is possible to give a topological characterization of holomorphically linearizable local dynamical systems:

Proposition 5.1. *Let $f \in \mathrm{End}\,(\mathbb{C}, 0)$ be a holomorphic local dynamical system with multiplier $0 < |\lambda| \leq 1$. Then f is holomorphically linearizable if and only if it is topologically linearizable if and only if 0 is contained in the interior of the stable set of f.*

Proof. If f is holomorphically linearizable, it is topologically linearizable and if it is topologically linearizable (and $|\lambda| \leq 1$) then K_f is an open neighbourhood of the origin. Assume then that 0 is contained in the interior of the stable set. If $0 < |\lambda| < 1$, we already know that f is holomorphically linearizable. If $|\lambda| = 1$, set

$$\varphi_k(z) = \frac{1}{k} \sum_{j=0}^{k-1} \frac{f^j(z)}{\lambda^j}$$

so that $\varphi_k'(0) = 1$ and

$$\varphi_k \circ f = \lambda \varphi_k + \frac{\lambda}{k} \left(\frac{f^k}{\lambda^k} - \mathrm{id} \right). \tag{21}$$

The hypothesis implies that there are bounded open sets $V \subset U$ containing the origin such that $f^k(V) \subset U$ for all $k \in \mathbb{N}$. Since $|\lambda| = 1$, it follows that $\{\varphi_k\}$ is a uniformly bounded family on V and hence, by Montel's theorem, it admits a converging subsequence. But (21) implies that a converging subsequence converges to a conjugation between f and the rotation $z \mapsto \lambda z$ and thus, f is holomorphically linearizable. □

The second important observation is that two elliptic holomorphic local dynamical systems with the same multiplier are always formally linearizable:

Proposition 5.2. *Let $f \in \mathrm{End}\,(\mathbb{C}, 0)$ be a holomorphic local dynamical system of multiplier $\lambda = e^{2\pi i\theta} \in S^1$ with $\theta \notin \mathbb{Q}$. Then f is formally conjugated to its linear part.*

Proof. We shall prove that there is a unique formal power series

$$h(z) = z + h_2 z^2 + \cdots \in \mathbb{C}[[z]]$$

such that $h(\lambda z) = f(h(z))$. Indeed, we have

$$h(\lambda z) - f(h(z)) =$$

$$= \sum_{j\geq 2} \left\{ \left[(\lambda^j - \lambda)h_j - a_j\right] z^j - \sum_{\ell=1}^{j} \binom{j}{\ell} z^{\ell+j} \left(\sum_{k\geq 2} h_k z^{k-2} \right)^{\ell} \right\}$$

$$= \sum_{j\geq 2} \left[(\lambda^j - \lambda)h_j - a_j - X_j(h_2, \ldots, h_{j-1})\right] z^j,$$

where X_j is a polynomial in $j-2$ variables. It follows that the coefficients of h are uniquely determined by induction using the formula

$$h_j = \frac{a_j + X_j(h_2, \ldots, h_{j-1})}{\lambda^j - \lambda}, \tag{22}$$

where X_j is a polynomial. In particular, h_j depends only on $\lambda, a_2, \ldots, a_j$. □

The formal power series linearizing f is not converging if its coefficients grow too fast. Thus, (22) links the radius of convergence of h to the behaviour of $\lambda^j - \lambda$: if the latter becomes too small, the series defining h does not converge. This is known as the *small denominators problem* in this context.

It is then natural to introduce the following quantity:

$$\Omega_\lambda(m) = \min_{1\leq k\leq m} |\lambda^k - 1|$$

for $\lambda \in S^1$ and $m \geq 1$. Clearly, λ is a root of unity if and only if $\Omega_\lambda(m) = 0$ for all m greater or equal to some $m_0 \geq 1$; furthermore,

$$\lim_{m\to+\infty} \Omega_\lambda(m) = 0$$

for all $\lambda \in S^1$.

The first one to prove that there are non-linearizable elliptic holomorphic local dynamical systems has been Cremer in 1927 ([8]). His more general result is the following:

Theorem 5.1 (Cremer, 1938 [9]). *Let $\lambda \in S^1$ be such that*

$$\limsup_{m\to+\infty} \frac{1}{m} \log \frac{1}{\Omega_\lambda(m)} = +\infty. \tag{23}$$

Then there exists $f \in \mathrm{End}\,(\mathbb{C}, 0)$ with multiplier λ which is not holomorphically linearizable. Furthermore, the set of $\lambda \in S^1$ satisfying (23) contains a G_δ-dense set.

Proof. Choose inductively $a_j \in \{0,1\}$ so that $|a_j + X_j| \geq 1/2$ for all $j \geq 2$, where X_j is as in (22). Then

$$f(z) = \lambda z + a_2 z^2 + \cdots \in \mathbb{C}_0\{z\},$$

while (23) implies that the radius of convergence of the formal linearization h is 0 and thus, f cannot be holomorphically linearizable, as required.

Let now $S(q_0) \subset S^1$ denote the set of $\lambda = e^{2\pi i\theta} \in S^1$ such that

$$\left|\theta - \frac{p}{q}\right| < \frac{1}{2^{q!}} \tag{24}$$

for some $p/q \in \mathbb{Q}$ in lowest terms with $q \geq q_0$. Then it is not difficult to check that each $S(q_0)$ is a dense open set in S^1 and that all $\lambda \in \mathcal{S} = \bigcap_{q_0 \geq 1} S(q_0)$ satisfy (23). Indeed, if $\lambda = e^{2\pi i\theta} \in \mathcal{S}$, we can find $q \in \mathbb{N}$ arbitrarily large such that there is $p \in \mathbb{N}$ so that (24) holds. Now, it is easy to see that

$$|e^{2\pi it} - 1| \leq 2\pi|t|$$

for all $t \in [-1/2, 1/2]$. Then let p_0 be the integer closest to $q\theta$ so that $|q\theta - p_0| \leq 1/2$. Then we have

$$|\lambda^q - 1| = |e^{2\pi iq\theta} - e^{2\pi ip_0}| = |e^{2\pi i(q\theta - p_0)} - 1| \leq 2\pi|q\theta - p_0| \leq 2\pi|q\theta - p| < \frac{4\pi}{2^{q!}}$$

for arbitrarily large q and (23) follows. □

On the other hand, Siegel gave a condition on the multiplier ensuring holomorphic linearizability in 1942:

Theorem 5.2 (Siegel, 1942 [35]). *Let $\lambda \in S^1$ be such that there exist $\beta > 1$ and $\gamma > 0$ such that*

$$\frac{1}{\Omega_\lambda(m)} \leq \gamma m^\beta \qquad \forall m \geq 2. \tag{25}$$

Then all $f \in \mathrm{End}\,(\mathbb{C}, 0)$ with multiplier λ are holomorphically linearizable. Furthermore, the set of $\lambda \in S^1$ satisfying (25) for some $\beta \geq 1$ and $\gamma > 0$ is of full Lebesgue measure in S^1.

Remark 5.1. It is interesting to notice that for generic (in a topological sense) $\lambda \in S^1$, there is a non-linearizable holomorphic local dynamical system with multiplier λ, while, for almost all (in a measure–theoretic sense) $\lambda \in S^1$, every holomorphic local dynamical system with multiplier λ is holomorphically linearizable.

The original proof of Theorem 5.2 was based on the method of majorant series, that requires finding a convergent series whose coefficients are greater than the coefficients of the formal linearization. A different proof is in the spirit of the so-called Kolmogorov–Arnold–Moser (or KAM) method (see [15]). Unfortunately, both proofs (as well as the proofs of the next two theorems) are well beyond the scope of this survey.

A bit of terminology is now useful: if $f \in \mathrm{End}\,(\mathbb{C},0)$ is elliptic, we shall say that the origin is a *Siegel point* if f is holomorphically linearizable; otherwise, it is a *Cremer point.*

Theorem 5.2 suggests the existence of a number–theoretical condition on λ ensuring that the origin is a Siegel point for any holomorphic local dynamical system of multiplier λ. And, indeed, this is the content of the celebrated *Bryuno–Yoccoz theorem*:

Theorem 5.3 (Bryuno, 1965 [3–5]; Yoccoz, 1988 [37–38]). *Let $\lambda \in S^1$. Then the following statements are equivalent:*

(i) the origin is a Siegel point for the quadratic polynomial $f_\lambda(z) = \lambda z + z^2$;
(ii) the origin is a Siegel point for all $f \in \mathrm{End}\,(\mathbb{C},0)$ with multiplier λ;
(iii) the number λ satisfies Bryuno's condition

$$\sum_{k=0}^{+\infty} \frac{1}{2^k} \log \frac{1}{\Omega_\lambda(2^{k+1})} < +\infty. \tag{26}$$

Bryuno, using majorant series as in Siegel's proof of Theorem 5.2 (see also [16] and references therein), has proved that condition (iii) implies condition (ii). Yoccoz, using a more geometric approach based on conformal and quasi–conformal geometry, has proved that (i) is equivalent to (ii) and that (ii) implies (iii), that is, that if λ does not satisfy (26) then the origin is a Cremer point for some $f \in \mathrm{End}\,(\mathbb{C},0)$ with multiplier λ — and hence, it is a Cremer point for the quadratic polynomial $f_\lambda(z)$. See also [32,33] for related results.

Remark 5.2. Conditions (23), (25) and (26) are usually expressed in a different way. Write $\lambda = e^{2\pi i\theta}$ and let $\{p_k/q_k\}$ be the sequence of rational numbers converging to θ given by the expansion in continued fractions. Then (26) is equivalent to

$$\sum_{k=0}^{+\infty} \frac{1}{q_k} \log q_{k+1} < +\infty,$$

while (25) is equivalent to

$$q_{n+1} = O(q_n^\beta)$$

and (23) is equivalent to

$$\limsup_{k\to+\infty} \frac{1}{q_k} \log q_{k+1} = +\infty.$$

See [16,38] and references therein for details.

If 0 is a Siegel point for $f \in \mathrm{End}\,(\mathbb{C}, 0)$, the local dynamics of f is completely clear and simple enough. On the other hand, if 0 is a Cremer point of f then the local dynamics of f is very complicated and not yet completely understood. Pérez-Marco ([26, 28–31]) and Biswas ([1]) have studied the topology and the dynamics of the stable set in this case. Some of their results are summarized in the following

Theorem 5.4 (Pérez-Marco, 1995 [30,31]; Biswas, 2007 [1]). *Assume that* 0 *is a Cremer point for an elliptic holomorphic local dynamical system* $f \in \mathrm{End}\,(\mathbb{C}, 0)$. *Then:*

(i) The stable set K_f *is compact, connected, full (i.e.,* $\mathbb{C}\backslash K_f$ *is connected), it is not reduced to* $\{0\}$ *and it is not locally connected at any point distinct from the origin.*
(ii) Any point of $K_f \setminus \{0\}$ *is recurrent (that is, a limit point of its orbit).*
(iii) There is an orbit in K_f *which accumulates at the origin, but no non-trivial orbit converges to the origin.*
(iv) The rotation number and the conformal class of K_f *are a complete set of holomorphic invariants for Cremer points. In other words, two elliptic non-linearizable holomorphic local dynamical systems* f *and* g *are holomorphically locally conjugated if and only if they have the same rotation number and there is a biholomorphism of a neighbourhood of* K_f *with a neighbourhood of* K_g.

Remark 5.3. So, if $\lambda \in S^1$ is not a root of unity and does not satisfy Bryuno's condition (26), we can find f_1, $f_2 \in \mathrm{End}\,(\mathbb{C}, 0)$ with multiplier λ such that f_1 is holomorphically linearizable while f_2 is not. Then f_1 and f_2 are formally conjugated without being neither holomorphically nor topologically locally conjugated.

See also [25,27] for other results on the dynamics about a Cremer point.

References

1. K. Biswas, *Complete conjugacy invariants of nonlinearizable holomorphic dynamics*, preprint (2007).
2. L. E. Böttcher, *The principal laws of convergence of iterates and their application to analysis*, Izv. Kazan. Fiz.-Mat. Obshch. **14** (1904), 155–234.
3. A. D. Bryuno, *Convergence of transformations of differential equations to normal forms*, Dokl. Akad. Nauk. USSR **165** (1965), 987–989.
4. A. D. Bryuno, *Analytical form of differential equations, I*, Trans. Moscow Math. Soc. **25** (1971), 131–288.
5. A. D. Bryuno, *Analytical form of differential equations, II*, Trans. Moscow Math. Soc. **26** (1972), 199–239.
6. C. Camacho, *On the local structure of conformal mappings and holomorphic vector fields*, Astérisque **59-60** (1978) 83–94.
7. S. Carleson and F. Gamelin, *Complex dynamics*, Springer, Berlin (1994).
8. H. Cremer, *Zum Zentrumproblem*, Math. An. **98** (1927), 151–163.
9. H. Cremer, *Über die Häufigkeit der Nichtzentren*, Math. Ann. **115** (1938), 573–580.
10. J. Écalle, *Les fonctions résurgentes. Tome I: Les algèbres de fonctions résurgentes*, Publ. Math. Orsay **81-05**, Université de Paris–Sud, Orsay (1981).
11. J. Écalle, *Les fonctions résurgentes. Tome II: Les fonctions résurgentes appliquées à l'itération*, Publ. Math. Orsay **81-06**, Université de Paris–Sud, Orsay (1981).
12. P. Fatou, *Sur les équations fonctionnelles, I*, Bull. Soc. Math. France **47** (1919), 161–271.
13. P. Fatou, *Sur les équations fonctionnelles, II*, Bull. Soc. Math. France **48** (1920), 33–94.
14. P. Fatou, *Sur les équations fonctionnelles, III*, Bull. Soc. Math. France **48** (1920), 208–314.
15. B. Hasselblatt and A. Katok, *Introduction to the modern theory of dynamical systems*, Cambridge Univ. Press, Cambridge (1995).
16. M. Herman, *Recent results and some open questions on Siegel's linearization theorem of germs of complex analytic diffeomorphisms of $\mathbb{C}^n$ near a fixed point*, Proc. 8th Int. Cong. Math. Phys., World Scientific, Singapore (1986), 138–198.
17. Y. S. Il'yashenko, *Nonlinear Stokes phenomena*, Adv. in Soviet Math. **14**, Am. Math. Soc., Providence (1993), 1–55.
18. T. Kimura, *On the iteration of analytic functions*, Funk. Eqvacioj **14** (1971), 197–238.
19. G. Kœnigs, *Recherches sur les integrals de certain equations fonctionelles*, Ann. Sci. Éc. Norm. Sup. **1** (1884) 1–41.
20. L. Leau, *Étude sur les equations fonctionelles à une ou plusieurs variables*, Ann. Fac. Sci. Toulouse **11** (1897), E1–E110.
21. B. Malgrange, *Travaux d'Écalle et de Martinet-Ramis sur les systèmes dynamiques*, Astérisque **92-93** (1981-82), 59–73.

22. B. Malgrange, *Introduction aux travaux de J. Écalle*, Ens. Math. **31** (1985), 261–282.
23. S. Marmi, *An introduction to small divisors problems*, I.E.P.I., Pisa (2000).
24. J. Milnor, *Dynamics in one complex variable*, Vieweg, Braunschweig (2000).
25. R. Pérez–Marco, *Sur les dynamiques holomorphes non linéarisables et une conjecture de V.I. Arnold*, Ann. Sci. École Norm. Sup. **26** (1993), 565–644.
26. R. Pérez–Marco, *Topology of Julia sets and hedgehogs* preprint (1994).
27. R. Pérez–Marco, *Non-linearizable holomorphic dynamics having an uncountable number of symmetries*, Invent. Math. **199** (1995), 67–127.
28. R. Pérez–Marco, *Holomorphic germs of quadratic type*, preprint (1995).
29. R. Pérez–Marco, *Hedgehogs dynamics*, preprint (1995).
30. R. Pérez–Marco, *Sur une question de Dulac et Fatou*, C.R. Acad. Sci. Paris **321** (1995), 1045–1048.
31. R. Pérez–Marco, *Fixed points and circle maps*, Acta Math. **179** (1997), 243–294.
32. R. Pérez–Marco, *Linearization of holomorphic germs with resonant linear part*, preprint (2000).
33. R. Pérez–Marco, *Total convergence or general divergence in small divisors*, Comm. Math. Phys. **223** (2001), 451–464.
34. A. A. Shcherbakov, *Topological classification of germs of conformal mappings with identity linear part*, Moscow Univ. Math. Bull. **37** (1982), 60–65.
35. C. L. Siegel, *Iteration of analytic functions*, Ann. of Math. **43** (1942), 607–612.
36. S. M. Voronin, *Analytic classification of germs of conformal maps* $(\mathbb{C},0) \to (\mathbb{C},0)$ *with identity linear part*, Func. Anal. Appl. **15** (1981), 1–17.
37. J.-C. Yoccoz, *Linéarisation des germes de difféomorphismes holomorphes de* $(\mathbb{C},0)$, C.R. Acad. Sci. Paris **306** (1988), 55–58.
38. J.-C. Yoccoz, *Théorème de Siegel, nombres de Bryuno et polynômes quadratiques*, Astérisque **231** (1995), 3–88.

NOTES IN TRANSFERENCE OF BILINEAR MULTIPLIERS*

OSCAR BLASCO

Department of Mathematics
Universitat de Valencia
Burjassot 46100 (Valencia), Spain
E-mail: oscar.blasco@uv.es

Keywords: Bilinear Hilbert transform, bilinear maximal functions, transference methods, discretization.

1. Notation and preliminaries

These notes contain an extended version of the talk given in the III International Course of Mathematical Analysis held in La Rábida (Huelva, Spain) in September 2007 and they are based on results appeared in [1–4]. I would like to thank the organizers for the kind hospitality and their nice working atmosphere that all the participants (students and professors) enjoyed during our stay.

Let us start by recalling some classical operators whose bilinear formulation will be considered throughout the paper. Let $f : \mathbb{R} \to \mathbb{C}$ belong to the Schwarzt class and write

$$H(f)(x) = \lim_{\varepsilon \to 0} \int_{|y|>\varepsilon} \frac{f(x-y)}{y} dy$$

and

$$H^*(f)(x) = \sup_{\varepsilon>0} \left| \int_{|y|>\varepsilon} \frac{f(x-y)}{y} dy \right|,$$

for the *Hilbert* and *maximal Hilbert transform* respectively.

We also write

$$M(f)(x) = \sup_{\varepsilon>0} \frac{1}{2\varepsilon} \int_{|y|<\varepsilon} |f(x-y)| dy,$$

*Partially supported by Proyecto MTM 2005-08350.

for *Hardy–Littlewood maximal function* and

$$I_\alpha(f)(x) = \int_{\mathbb{R}} \frac{f(x-y)}{|y|^{1-\alpha}} dy,$$

for the *Fractional Integral* where $0 < \alpha < 1$.

They are very classical operators in Harmonic Analysis and are rather well understood not only in $\mathbb{R}$ but in many other groups and not only for the Lebesgue measure but for weight functions $w(x)dx$. Of course the boundedness in the setting of Lebesgue (and many other function spaces) of these operators (not entering in the extreme cases) is well known. Let recall that there exist constants $A_p, B_p, C_p, D_p > 0$ such that

$$\begin{aligned} &\|H(f)\|_p \leq A_p \|f\|_p, \\ &\|H^*(f)\|_p \leq B_p \|f\|_p \end{aligned} \qquad \text{for } 1 < p < \infty. \tag{1}$$

$$\|M(f)\|_p \leq C_p \|f\|_p, \qquad \text{for } 1 < p \leq \infty. \tag{2}$$

$$\|I_\alpha(f)\|_q \leq D_p \|f\|_p, \qquad \text{for } 0 < \alpha < \frac{1}{p},\ 1 < p < \infty,\ \frac{1}{q} = \frac{1}{p} - \alpha. \tag{3}$$

There are bilinear versions of these operators that have been studied in the last decade and which will be the aim of our considerations.

Given $f, g : \mathbb{R} \to \mathbb{C}$ belonging to the Schwarzt class we can now define the *bilinear Hilbert transform* by

$$H(f,g)(x) = \lim_{\varepsilon \to 0} \int_{|y|>\varepsilon} \frac{f(x-y)g(x+y)}{y} dy,$$

the *bisublinear maximal Hilbert transform* by

$$H^*(f,g)(x) = \sup_{\varepsilon>0} \left| \int_{|y|>\varepsilon} \frac{f(x-y)g(x+y)}{y} dy \right|,$$

the *bisublinear Hardy–Littlewood maximal function* by

$$M(f,g)(x) = \sup_{\varepsilon>0} \frac{1}{2\varepsilon} \int_{|y|<\varepsilon} |f(x-y)g(x+y)| dy,$$

and the *bilinear fractional integral* by

$$I_\alpha(f,g)(x) = \int_{\mathbb{R}} \frac{f(x-y)g(x+y)}{|y|^{1-\alpha}} dy, \quad 0 < \alpha < 1.$$

It has been the effort of several authors and many years to get the range of boundedness for the corresponding bilinear versions. We collect in the following theorem the actual knowledge of the problem.

Theorem 1.1. *Let $1 < p_1, p_2 < \infty$, $0 < \alpha < 1/p_1 + 1/p_2$, $1/q = 1/p_1 + 1/p_2 - \alpha$, $1/p_3 = 1/p_1 + 1/p_2$ and $2/3 < p_3 < \infty$. Then there exist constants A, B, C, D such that*

$$\|H(f,g)\|_{p_3} \leq A\|f\|_{p_1}\|g\|_{p_2} \text{ (Lacey–Thiele [16–18])}, \tag{4}$$

$$\|H^*(f,g)\|_{p_3} \leq B\|f\|_{p_1}\|g\|_{p_2} \text{ (Lacey [15])}, \tag{5}$$

$$\|M(f,g)\|_{p_3} \leq C\|f\|_{p_1}\|g\|_{p_2} \text{ (Lacey [15])}, \tag{6}$$

$$\|I_\alpha(f,g)\|_q \leq A\|f\|_{p_1}\|g\|_{p_2} \text{ (Kenig–Stein [14], Grafakos–Kalton [13])}. \tag{7}$$

We would like to consider analogue operators in the periodic or the discrete case and to analyze their boundedness.

In particular, one can define the *bilinear conjugate function* as

$$B(F,G)(e^{it}) = \int_{-\pi}^{\pi} F(t-s)G(t+s)\cot(s/2)\frac{ds}{2\pi}$$

where F and G are polynomials on $\mathbb{T}$.

Using Fourier series expansion of the polynomials, the operator can also be written as

$$B(F,G)(e^{it}) = -i\sum_k \left(\sum_{n+m=k} \text{sign } (n-m)\hat{F}(n)\hat{G}(m) \right) e^{ikt}$$

where $F(t) = \sum_{-N}^{N} \hat{F}(n)e^{int}$ and $G(t) = \sum_{-M}^{M} \hat{G}(m)e^{imt}$.

The fundamental question is the following: *Is the bilinear conjugate transform bounded from $L^{p_1}(\mathbb{T}) \times L^{p_2}(\mathbb{T}) \to L^{p_3}(\mathbb{T})$ for some values of p_1, p_2, p_3?*

While the situation in the linear case reduces to adapt the proof of the group $\mathbb{R}$ to the group $\mathbb{T}$ (or to replace the half–space for the disc when using a complex–variable approach), the techniques that were needed for the real line in the bilinear case do not seem to have any easy modification to the periodic setting to obtain the boundedness of the bilinear conjugate function defined in $\mathbb{T}$. However some transference techniques known in the linear case can be adapted to the bilinear one.

Another analogue formulations that we would like to consider are *discrete bilinear Hilbert transform*, the *discrete bisublinear Hardy-Littlewood maximal function* and the *discrete bilinear fractional transform*, defined by

$$H_d(\lambda,\beta)(n) = \lim_{N\to\infty} \sum_{0<|k|\leq N} \frac{\lambda_{n-k}\beta_{n+k}}{k},$$

$$M_d(\lambda,\beta)(n) = \sup_{N\in\mathbb{N}} \frac{1}{2N} \sum_{0<|k|\leq N} |\lambda_{n-k}||\beta_{n+k}| \text{ and}$$

$$I_d^\alpha(\lambda,\beta)(n) = \sum_{k\neq 0, k\in\mathbb{Z}} \frac{\lambda_{n-k}\beta_{n+k}}{|k|^{1-\alpha}}$$

for finite sequences λ, β respectively.

As above, the fundamental question is the following: *Are they bounded from $\ell^{p_1}(\mathbb{Z}) \times \ell^{p_2}(\mathbb{Z}) \to \ell^{p_3}(\mathbb{Z})$ for some values of p_1, p_2, p_3?*

Several methods have been developed to such purposes in the last five years. In fact two different approaches have been applied. The first one is the bilinear formulation of the DeLeeuw method [8] first considered in the paper by Fan and Sato [9] and then developed by O. Blasco and P. Villarroya [1,5]. The second one is the bilinear formulation of the Coifman–Weiss transference method [7] that has been extensively studied in [2–4] by O. Blasco, E. Berkson, M. J. Carro and A.T. Gillespie.

We shall only mention one theorem and its application of each of the procedures considered in the just mentioned papers. All the results appearing in Theorem 1.1 can be transferred to both situations periodic and discrete. We will also present a detailed proof for the reader to see the tools used in our approaches. The interested reader can consult the references in the bibliography for a further study of the topic.

2. Methods and applications

Let us start considering the simplest situation, corresponding to bilinear convolution with integrable kernels.

Assume $K \in L^1(\mathbb{R})$ and define

$$B_K(f,g)(x) = \int_{\mathbb{R}} f(x-y)g(x+y)K(y)dy.$$

Writing $f(x-y) = \int_{\mathbb{R}} \hat{f}(\xi)e^{i(x-y)\xi}d\xi$ and $g(x+y) = \int_{\mathbb{R}} \hat{g}(\eta)e^{i(x+y)\eta}d\eta$, we can also use expression:

$$\begin{aligned}
B_K(f,g)(x) &= \int_{\mathbb{R}} f(x-y)g(x+y)K(y)dy \\
&= \int_{\mathbb{R}}\int_{\mathbb{R}}\int_{\mathbb{R}} \hat{f}(\xi)\hat{g}(\eta)K(y)e^{i(x-y)\xi}e^{i(x+y)\eta}d\xi d\eta dy \\
&= \int_{\mathbb{R}}\int_{\mathbb{R}} \hat{f}(\xi)\hat{g}(\eta)\left(\int_{\mathbb{R}} K(y)e^{-i(\xi-\eta)y}dy\right)e^{i(\xi+\eta)x}d\xi d\eta \\
&= \int_{\mathbb{R}}\int_{\mathbb{R}} \hat{g}(\eta)\hat{f}(\xi)\hat{K}(\xi-\eta)e^{i(\xi+\eta)x}d\xi d\eta.
\end{aligned}$$

This motivates the following definition.

Definition 2.1. Let $0 < p_1, p_2, p_3 < \infty$ and $1/p_1 + 1/p_2 = 1/p_3$. A bounded measurable function $m(\xi, \eta)$ is said to be a *bilinear multiplier* on $\mathbb{R}$ of type (p_1, p_2, p_3) if the operator

$$B_m(f,g)(x) = \int_{\mathbb{R}} \int_{\mathbb{R}} \hat{f}(\xi)\hat{g}(\eta)m(\xi,\eta)e^{i(\xi+\eta)x}\,d\xi d\eta$$

is bounded from $L^{p_1}(\mathbb{R}) \times L^{p_2}(\mathbb{R})$ to $L^{p_3}(\mathbb{R})$.

The study of bilinear multipliers for smooth symbols (where $m(\xi, \eta)$ is a “nice” regular function) goes back to the work by R. R. Coifman and Y. Meyer in [6].

Let us restrict ourselves to a smaller family of multipliers: the case $m(\xi, \eta) = m'(\xi - \eta)$ where $m'(x)$ is bounded in $\mathbb{R}$. The simplest case is $m'(x) = \hat{\mu}(x)$ where μ is a Borel regular measure in $\mathbb{R}$. It is elementary to see that m' define a bilinear multiplier on $\mathbb{R}$ of type (p_1, p_2, p_3) whenever $p_3 \geq 1$ and $1/p_1 + 1/p_2 = 1/p_3$. Indeed, using the expression

$$B_m(f,g)(x) = \int_{\mathbb{R}} f(x-t)g(x+t)d\mu(t)$$

one gets

$$\begin{aligned}
\|B_m(f,g)\|_{p_3} &\leq \int_{\mathbb{R}} \|f(\cdot - t)g(\cdot + t)\|_{p_3} d|\mu|(t) \\
&\leq \int_{\mathbb{R}} \|f(\cdot - t)\|_{p_1} \|g(\cdot + t)\|_{p_2} d|\mu|(t) \\
&= \|f\|_{p_1}\|g\|_{p_2} \int_{\mathbb{R}} d|\mu|(t) = \|\mu\|_1 \|f\|_{p_1}\|g\|_{p_2}.
\end{aligned}$$

However, the case where the symbol m' is not smooth has a much shorter story.

A very non trivial example is given by $m'(x) = -i\,\text{sign}\,(x)$ which leads to the bilinear Hilbert transform and it was first considered by Lacey and Thiele in [16–18] and then extended to other cases in [10,11]. The solution took many years to be achieved after the formulation of the question by A. P. Calderón in the seventies.

Let us mention a general method to transfer results from $\mathbb{R}$ to $\mathbb{T}$. The approach follows the DeLeeuw method in the linear case and there are two different proofs of the following result.

Theorem 2.1 ([1,9]). *Let $m(\xi, \eta)$ be a continuous function defining a bilinear multiplier on $\mathbb{R}$ of type (p_1, p_2, p_3) where $1/p_1 + 1/p_2 = 1/p_3$ and*

$p_3 \geq 1$, *i.e., the operator*

$$B_m(f,g)(x) = \int_{\mathbb{R}} \int_{\mathbb{R}} \hat{f}(\xi)\hat{g}(\eta)m(\xi,\eta)e^{i(\xi+\eta)x} d\xi d\eta$$

is bounded from $L^{p_1}(\mathbb{R}) \times L^{p_2}(\mathbb{R})$ *to* $L^{p_3}(\mathbb{R})$. *Then the sequence* $m_{k,k'} = m(k,k')$ *define a bilinear multiplier on* $\mathbb{T}$ *of type* (p_1,p_2,p_3), *i.e., the operator*

$$\tilde{B}_m(F,G)(t) = \sum_k \Big(\sum_{n+n'=k} \hat{F}(n)\hat{G}(n')m(n,n') \Big) e^{ikt}$$

is bounded from $L^{p_1}(\mathbb{T}) \times L^{p_2}(\mathbb{T})$ *to* $L^{p_3}(\mathbb{T})$.

Corollary 2.1. *The bilinear conjugate function operator is bounded from* $L^{p_1}(\mathbb{T}) \times L^{p_2}(\mathbb{T})$ *to* $L^{p_3}(\mathbb{T})$ *whenever* $1 < p_1, p_2 < \infty$, $1/p_1 + 1/p_2 = 1/p_3$ *and* $p_3 \geq 1$.

The reader should be aware that the restriction $p_3 \geq 1$ is a limitation of the proof but it can be removed using other approaches (see [4]).

To handle the discrete case, there are also two different techniques (see [1] or [2,4]). We shall select here the second approach using a "discretization" method.

Let us define the mappings $P : \ell^p(\mathbb{Z}) \to L^p(\mathbb{R})$ by

$$\lambda = (\lambda_n) \to f = \sum_{n\in\mathbb{Z}} \lambda_n \chi_{(n-1/4,n+1/4)}$$

and $Q : L^p(\mathbb{R}) \to \ell^p(\mathbb{Z})$ by

$$f \to \left(\int_{(n-1/4,n+1/4)} f(x)dx \right)_{n\in\mathbb{Z}}.$$

Clearly $\|P(\lambda)\|_p = C\|\lambda\|_p$ for $0 < p < \infty$ and $\|Q(f)\|_p \leq C\|f\|_p$ for $1 \leq p < \infty$.

Theorem 2.2 ([4]). *Let* K *be integrable in* $\mathbb{R}$ *and denote*

$$B_K(f,g)(x) = \int_{\mathbb{R}} f(x-y)g(x+y)K(y)dy$$

for f *and* g *simple functions. If*

$$K_n = \int_{[-1/4,1/4]} \int_{[n-1/4,n+1/4]} K(x-u)K(x+u)dxdy$$

then

$$QB_K\Big(P(\lambda),P(\beta)\Big)(n) = \sum_{k\in\mathbb{Z}} \lambda_{n-k}\beta_{n+k}K_k$$

for any finite sequences λ and β.

In particular, for $p_3 \geq 1$ one has that QB_KP is bounded from $\ell^{p_1}(\mathbb{Z}) \times \ell^{p_2}(\mathbb{Z})$ to $\ell^{p_3}(\mathbb{Z})$ with norm bounded by the norm of B_K as an operator from $L^{p_1}(\mathbb{R}) \times L^{p_2}(\mathbb{R})$ to $L^{p_3}(\mathbb{R})$).

Corollary 2.2. *The bilinear discrete Hilbert transform is bounded from $\ell^{p_1}(\mathbb{Z}) \times \ell^{p_2}(\mathbb{Z})$ to $\ell^{p_3}(\mathbb{Z})$ whenever $1 < p_1, p_2 < \infty$, $1/p_1 + 1/p_2 = 1/p_3$ and $p_3 \geq 1$.*

The reader should also be aware that the restriction $p_3 \geq 1$ is again a limitation of the proof but it was removed in [2] to get $p_3 > \frac{2}{3}$.

Let us finally explain a bit how to get the transference method of Coifman–Weiss in the bilinear setting (see [2–4]).

Let G be a l.c.a group with Haar measure m, let (Ω, Σ, μ) be a measure space and let R_u be a representation of G in the space of bounded linear operators on $L^p(\mu)$, i.e., $R : G \to L(L^p(\mu), L^p(\mu))$ such that $u \to R_u$ verifies

- $R_u R_v = R_{uv}$ for $u, v \in G$,
- $\lim_{u \to 0} R_u f = f$ for $f \in L^p(\mu)$,
- $\sup_{u \in G} \|R_u\| < \infty$.

Let $K \in L^1(G)$ with compact support. Denote now

$$B_K(\phi, \psi)(v) = \int_G \phi(v-u)\psi(v+u)K(u)dm(u)$$

for ϕ, ψ simple functions defined on G and assume that, for $0 < p_1, p_2 < \infty$ and $1/p_1 + 1/p_2 = 1/p_3$, the bilinear operator B_K is bounded from $L^{p_1}(G) \times L^{p_2}(G)$ to $L^{p_3}(G)$ with "norm" $N_{p_1,p_2}(B_K)$.

We now consider the transferred operator by the formula

$$T_K(f, g)(w) = \int_G R_{-u}f(w)R_u g(w)K(u)dm(u)$$

for $f \in L^{p_1}(\mu)$ and $g \in L^{p_2}(\mu)$.

Let us present, in a particular case, a prototype result that one can produce in this setting. The assumptions can be weakened and the setting can be relaxed but we concentrate in the case for simplicity.

Theorem 2.3. *Let $G = \mathbb{R}$, (Ω, Σ, μ) a measure space, $1 \leq p_1, p_2 < \infty$ and $1/p_3 = 1/p_1 + 1/p_2$. Let R be a representation of $\mathbb{R}$ on acting $L^{p_i}(\mu)$ for $i = 1, 2$ with*

$$\sup_{u \in \mathbb{R}} \|R_u\|_{L(L^{p_i}, L^{p_i})} = 1$$

for $i = 1, 2$.

Assume that there exists a map $u \to L(L^{p_3}(\mu), L^{p_3}(\mu))$ *given by* $u \to S_u$ *such that* S_u *are invertible with* $\sup_{u\in G} \|S_u^{-1}\| = 1$ *and*

$$S_v((R_{-u}f)(R_u g)) = (R_{v-u}f)(R_{v+u}g)$$

for $u, v \in \mathbb{R}$, $f \in L^{p_1}(\mu)$ *and* $g \in L^{p_2}(\mu)$.

Let K *belong to* $L^1(\mathbb{R})$ *and be supported in* $[-A, A]$. *If the bilinear map* B_K *defined as above is bounded with norm* $N_{p_1,p_2}(B_K)$ *then* T_K *is also bounded from* $L^{p_1}(\mu) \times L^{p_2}(\mu)$ *to* $L^{p_3}(\mu)$ *and with norm bounded by* $CN_{p_1,p_2}(B_K)$.

Proof. Write, for each $v \in \mathbb{R}$,

$$\begin{aligned} T_K(f,g) &= S_v^{-1}\left(S_v \int_{\mathbb{R}} R_{-u}fR_u gK(u)du\right) \\ &= S_v^{-1}\left(\int_{\mathbb{R}} S_v(R_{-u}fR_u g)K(u)du\right) \\ &= S_v^{-1}\left(\int_{\mathbb{R}} (R_{v-u}f)(R_{v+u}g)K(u)du\right) \end{aligned}$$

Hence

$$\|T_K(f,g)\|_{L^{p_3}(\mu)}^{p_3} \leq \left\| \int_{\mathbb{R}} (R_{v-u}f)(R_{v+u}g)K(u)du \right\|_{L^{p_3}(\mu)}^{p_3}.$$

Given $N \in \mathbb{N}$, integrating over $v \in [-N, N]$,

$$2N\|T_K(f,g)\|_{L^{p_3}(\mu)}^{p_3} \leq \int_{-N}^{N} \left\| \int_{\mathbb{R}} (R_{v-u}f)(R_{v+u}g)K(u)du \right\|_{L^{p_3}(\mu)}^{p_3} dm(v).$$

Therefore

$$2N\,\|T_K(f,g)\|^{p_3}_{L^{p_3}(\mu)} \le \int_{-N}^{N}\int_\Omega \left|\int_{\mathbb{R}} R_{v-u}f(w)R_{v+u}g(w)K(u)du\right|^{p_3} d\mu(w)dv$$

$$= \int_\Omega \left(\int_{-N}^{N}\left|\int_{-A}^{A} R_{v-u}f(w)R_{v+u}g(w)K(u)du\right|^{p_3} dv\right) d\mu(w)$$

$$= \int_\Omega \left(\int_{\mathbb{R}} |B_K(R_uf(w)\chi_{[-A-N,A+N]}, R_ug(w)\chi_{[-A-N,A+N]})(v)|^{p_3}\, dv\right)$$

$$\times\ d\mu(w)$$

$$= \int_\Omega \|B_K(R_uf(w)\chi_{[-A-N,A+N]}, R_ug(w)\chi_{[-A-N,A+N]})\|^{p_3}_{L^{p_3}(\mathbb{R})}\, d\mu(w)$$

$$\le\ N_{p_1,p_2}(B_K)^{p_3}\left(\int_\Omega \|R_uf(w)\chi_{[-A-N,A+N]}\|^{p_1}_{L^{p_1}(\mathbb{R})}\, d\mu(w)\right)^{p_3/p_1}$$

$$\times\left(\int_\Omega \|R_ug(w)\chi_{[-A-N,A+N]}\|^{p_2}_{L^{p_2}(\mathbb{R})}\, d\mu(w)\right)^{p_3/p_2}$$

$$= N_{p_1,p_2}(B_K)^{p_3}\left(\int_{-(A+N)}^{A+N} \|R_uf\|^{p_1}_{L^{p_1}(\mu)}\, du\right)^{p_3/p_1}$$

$$\times\left(\int_{-(A+N)}^{A+N} \|R_ug\|^{p_2}_{L^{p_2}(\mu)}\, du\right)^{p_3/p_2}$$

$$= N_{p_1,p_2}(B_K)^{p_3}\left(\int_{-(A+N)}^{A+N} \|f\|^{p_1}_{L^{p_1}(\mu)}\, du\right)^{p_3/p_1}$$

$$\times\left(\int_{-(A+N)}^{A+N} \|g\|^{p_2}_{L^{p_2}(\mu)}\, du\right)^{p_3/p_2}$$

$$\le\ N_{p_1,p_2}(B_K)^{p_3}(2(A+N))\|f\|^{p_3}_{L^{p_1}(\mu)}\|g\|^{p_3}_{L^{p_2}(\mu)}.$$

Therefore

$$\|T_K(f,g\|_{L^{p_3}(\mu)} \le \left(\frac{A+N}{N}\right)^{1/p_3} N_{p_1,p_2}(B_K)\|f\|^{p_3}_{L^{p_1}(\mu)}\|g\|^{p_3}_{L^{p_2}(\mu)}. \qquad \square$$

Note that, in particular, the assumptions in the previous theorem hold for multiplicative representations, i.e., $R_u(fg) = (R_uf)(R_ug)$, selecting $S_u = R_u$.

Let us finish with an application to Ergodic theory. We state here the result for maximal version of the operators, but results in the same spirit can be seen in [2,4]. Let (Ω, Σ, μ) be σ-finite measure space and T an invertible and bounded operator on $L^p(\mu)$. Define the *bisublinear maximal ergodic transform* by

$$M_T(f,g)(w) = \sup_{N>0} \frac{1}{2N} \sum_{n=-N}^{N} T^n f(w) T^{-n} g(w),$$

and the *bisublinear maximal ergodic Hilbert transform* by

$$H_T^*(f,g)(w) = \sup_{N>0} \sum_{0<|n|<N} \frac{T^n f(w) T^{-n} g(w)}{n}.$$

Theorem 2.4 ([2,3]). *Let $1 < p_1, p_2 < \infty$ and $1/p_1 + 1/p_2 = 1/p_3 < 3/2$, let T be an invertible operator on $L^{p_i}(\mu)$ for $i = 1,2$ such that T and T^{-1} are power bounded. Assume that there exists an invertible operator S defined on $L(L^{p_3}(\mu), L^{p_3}(\mu))$ such that*

$$S^m(T^n f T^{-n} g) = T^{m+n} f T^{m-n} g$$

for $f \in L^{p_1}(\mu)$ and $g \in L^{p_2}(\mu)$. Then M_T and H_T^* are bounded from $L^{p_1}(\mu) \times L^{p_2}(\mu)$ to $L^{p_3}(\mu)$.

References

1. O. Blasco, *Bilinear multipliers and transference*, Int. J. Math. Math. Sci. **2005**(4) (2005), 545–554.
2. E. Berkson, O. Blasco, M. Carro and A. Gillespie, *Discretization and transference of bisublinear maximal operators*, J. Fourier Anal. and Appl. **12** (2006), 447–481.
3. E. Berkson, O. Blasco, M. Carro and A. Gillespie, *Discretization versus transference for bilinear operators*, (2008), to appear.
4. O. Blasco, M. Carro and A. Gillespie, *Bilinear Hilbert transform on measure spaces*, J. Fourier Anal. and Appl. **11** (2005), 459–470.
5. O. Blasco and F. Villarroya, *Transference of bilinear multipliers on Lorentz spaces*, Illinois J. Math. **47**(4) (2005), 1327–1343.
6. R. R. Coifman and Y. Meyer, *Fourier Analysis of multilinear convolution, Calderón theorem and analysis of Lipschitz curves*, Euclidean Harmonic Analysis (Proc. Sem. Univ. Maryland, College Univ., Md) Lecture Notes in Math. **779** (1979), 104–122.
7. R. R. Coifman and W. Weiss, *Transference Methods in Analysis*, Regional Conf. Series in Math. **31**, Amer. Math. Soc., Providence (1977).
8. K. DeLeeuw, *On L_p-multipliers.*, Ann. Math. **91** (1965), 364–379.
9. D. Fan and S. Sato, *Transference of certain multilinear multipliers operators*, J. Austral. Math. Soc. **70** (2001), 37–55.
10. J. Gilbert and A. Nahmod, *Bilinear operators with non-smooth symbols*, J. Fourier Anal. Appl. **7** (2001), 435–467.

11. J. Gilbert and A. Nahmod, *Boundedness Bilinear operators with non-smooth symbols*, Mat. Res. Letters **7** (2000), 767–778.
12. L. Grafakos and P. Honzík, *Maximal transference and summability of multilinear Fourier series*, J. Australian Math. Soc. **80** (2006), 65–80.
13. L. Grafakos and N. Kalton, *Some remarks on multilinear maps and interpolation*, Math. Annalen **319** (2001), 151–180.
14. C. E. Kenig and E. M. Stein, *Multilinear estimates and fractional integration*, Math. Res. Lett. **6** (1999), 1–15.
15. M. Lacey, *The bilinear maximal function maps into L^p for $\frac{2}{3} < p \leq 1$* , Ann. Math. **151** (2000), 35–57.
16. M. Lacey and C. Thiele, *L^p estimates on the bilinear Hilbert transform for $2 < p < \infty$*, Annals Math. **146** (1997), 693–724.
17. M. Lacey and Thiele, C. *Weak bounds for the bilinear Hilbert transform on L^p*, Documenta Mathematica, extra volume ICM 1-1000, [1997].
18. M. Lacey and C. Thiele, *On Calderón's conjecture*, Ann. Math. **149**(2) (1999), 475–496.

GENERATING FUNCTIONS OF LEBESGUE SPACES BY TRANSLATIONS

JOAQUIM BRUNA*

Departament de Matemàtiques
Universitat Autònoma de Barcelona
08193 Barcelona, Spain
E-mail: bruna@mat.uab.cat

1. Discrete translates, generating functions

We will be using the following notations. For a discrete set $\Lambda \subset \mathbb{R}$ and $f \in L^p(\mathbb{R})$, $\Lambda(f)$ stands for the set of translates $\tau_\lambda f(x) = f(x-\lambda)$, $\lambda \in \Lambda$. It is a well–known result in the theory of frames that for $\varphi \in L^2(\mathbb{R})$, no $\Lambda(\varphi)$ is a frame in $L^2(\mathbb{R})$, that is, no φ, Λ can satisfy

$$A\|f\|^2 \leq \sum_{\lambda \in \Lambda} |\langle f, \tau_\lambda \varphi \rangle|^2 \leq B\|f\|^2$$

for some constants A, B. This amounts to say that, in order to have something useful for applications, translations alone are not enough and something else must be considered, either modulations (leading to time–frequency wavelets) or else dilations (leading to time–scale wavelets). From a pure mathematical viewpoint this fact raises the question to investigate what, if not frames, one can built with only translations. This is going to be the main theme in this lecture (related to the above, the following seems to be still an open question: can $\Lambda(\varphi)$ be a Schauder basis of $L^2(\mathbb{R})$?)

In case $\Lambda(\varphi)$ spans $L^p(\mathbb{R})$ (meaning that the space $T(\varphi, \Lambda)$ of linear combinations of functions in $\Lambda(\varphi)$ is dense) we say that φ is a Λ- generator of $L^p(\mathbb{R})$. We call φ an $L^p(\mathbb{R})$-generator if it is a Λ-generator for some Λ. Analogously, we say that Λ is a spectral set for $L^p(\mathbb{R})$ if some Λ-generator exists. The main issue in this talk is about characterizations of these concepts, and is based on the papers [5], [4] and [3].

*Supported by grants MTM2005-08984-C02-01 and 2005SGR00611.

Of course we need quoting first Wiener's and Beurling's theorems, dealing with all translates of φ, respectivelly in $L^1(\mathbb{R})$ and $L^2(\mathbb{R})$. As in all problems dealing with translations, an essential tool is the Fourier transform that we use here in the form $\hat{\varphi}(\xi) = \int \varphi(x)e^{-ix\xi}\,dx$. Wiener's theorem asserts that all translates of $\varphi \in L^1(\mathbb{R})$ span $L^1(\mathbb{R})$ if and only if $\hat{\varphi}(\xi) \neq 0$ for all ξ, while Beurling's theorem states that all translates of $\varphi \in L^2(\mathbb{R})$ span $L^2(\mathbb{R})$ if and only if $\hat{\varphi}(\xi) \neq 0$ a.e. Incidentally, it is to be noted that the analogous statement for $1 < p < 2$ seems to be unknown.

2. Connections with density of exponentials

Next we point out some known facts. We will be using the notation $\mathcal{E}(\Lambda)$ for the space of linear combinations of the characters $e^{i\xi\lambda}$, $\lambda \in \Lambda$ so that by Fourier transform $T(\varphi, \Lambda)$ corresponds to $\hat{\varphi}\mathcal{E}(\Lambda)$. By Parseval's theorem, φ is a $L^2(\mathbb{R})$-generator if and only if this later space is dense in $L^2(\mathbb{R})$, that is,

$$\int \left| g(\xi) - \hat{\varphi}(\xi) \sum a_\lambda e^{i\xi\lambda} \right|^2 \,d\xi$$

can be made arbitrarily small for any $g \in L^2(\mathbb{R})$. It is convenient to restate this in terms of weighted approximation as follows. We write $g = \hat{\varphi}h$, so h is a general function in the L^2-weighted space $L^2(\mathbb{R}, |\hat{\varphi}|^2)$ and the above is written

$$\int \left| h(\xi) - \sum a_\lambda e^{i\xi\lambda} \right|^2 |\hat{\varphi}(\xi)|^2 \,d\xi.$$

Thus we see that φ is a Λ-generator of $L^2(\mathbb{R})$ if and only if $\mathcal{E}(\Lambda)$ spans the weighted space $L^2(\mathbb{R}, |\hat{\varphi}|^2)$. For integer translates ($\Lambda = \mathbb{Z}$), it is clear that this can never be the case, because $\mathcal{E}(\mathbb{Z})$ consists of 2π periodic functions. A similar argument works in $L^p(\mathbb{R}), 1 \leq p \leq 2$. Surprisingly enough, Nikolskii ([9]) and later Atzmon–Olevskii ([1]) showed that for $p > 2$, $L^p(\mathbb{R})$ has $\mathbb{Z}$-generators. For close to integer translates, however, Olevskii ([11]) showed that $\Lambda = \{n + a_n \colon a_n \neq 0, a_n \to 0\}$ is a spectral set for $L^2(\mathbb{R})$; later, Olevski and Ulanowski ([12]) showed that if $|a_n| = O(r^{|n|})$ for some $r < 1$ then the generator φ can be chosen in the Schwartz class.

Suppose that φ is a Λ-generator of $L^2(\mathbb{R})$. If $E_{\varepsilon,N} = \{\xi \colon \varepsilon \leq |\hat{\varphi}(\xi)| \leq N\}$ then, since $\hat{\varphi} \neq 0$ a.e., $E_{\varepsilon,N}$ has arbitrarily large measure and $L^2(E_{\varepsilon,N}, |\hat{\varphi}|^2)$ is the usual space $L^2(E_{\varepsilon,N})$. Hence we see that if φ is a Λ-generator of $L^2(\mathbb{R})$ then $\mathcal{E}(\Lambda)$ is dense in $L^2(E)$ for sets E of arbitrarily large measure.

This shows the connection of these questions with the subject of *density of exponentials* $\mathcal{E}(\Lambda)$ in function spaces and, in particular, with Landau's results. Landau ([10]) constructed certain perturbations of the integers $\Lambda = \{n + a_n\}$ where a_n are bounded, such that $\mathcal{E}(\Lambda)$ is dense in L^2 on any finite union of the intervals $(2\pi(k-1)+\varepsilon, 2\pi k-\varepsilon)$, $\varepsilon > 0$, in particular on sets with arbitrarily large measure. In [13], Landau's result was extended to every sequence Λ as above, where a_n have an exponential decay. We mention here that if $\mathcal{E}(\Lambda)$ is complete in L^2 on "Landau sets", then one can construct a Λ-generator for $L^2(\mathbb{R})$ which belongs to the Schwartz class $S(\mathbb{R})$.

All the above shows as well that in the L^2 case it is hopeless to look for a characterization of the spectral sets in terms of some density of some kind, as $\mathbb{Z}$ and its perturbations would have the same density under any reasonable definition. This is so because the above shows that what is involved is the density of complex exponentials in $L^2(\mathbb{R}, \omega)$ where ω is positive a.e. but still might have some zeros, so that the sets $E_{\varepsilon,N}$ considered before might be disconnected, say unions of intervals. In such situation, questions about density of exponentials become very subtle and depend on arithmetic properties of the frequencies. However, the situation is greatly simplified when looking at the problem in $L^1(\mathbb{R})$, as we proceed to explain.

3. The problem in $L^1(\mathbb{R})$ and its connections with the spectral radious problem

Since the $L^1(\mathbb{R})$-norm of f dominates the sup–norm of $\hat{f}$, it is clear that if φ is an $L^1(\mathbb{R})$-generator then $\hat{\varphi}\mathcal{E}(\Lambda)$ would be dense in the sup–norm in any space X of functions included in the range of $L^1(\mathbb{R})$ by Fourier transform. The fact that now $\hat{\varphi}$ never vanishes makes a great difference; fix $\rho > 0$ and consider the interval $[-\rho, +\rho]$ on which $\hat{\varphi}$ is bounded above and below and so we may think it is one. Taking as X the space of test functions supported in $(-\rho, +\rho)$ we conclude that $\mathcal{E}(\Lambda)$ is dense in the sup–norm in X and hence in any reasonable function space on $[-\rho, +\rho]$, for all $\rho > 0$. This shows that what is involved in this case is the density of exponentials in *intervals*, and this is a much better known situation, classical in harmonic analysis.

We recall that the *spectral radious* of a set $\Lambda \subset \mathbb{R}$ is defined by

$$R(\Lambda) = \sup\{\rho > 0 \colon \mathcal{E}(\Lambda) \text{ is dense in } L^2([-\rho, \rho])\} \leq +\infty.$$

This means that $\mathcal{E}(\Lambda)$ is dense in $L^2([-\rho, +\rho])$ if $\rho < R(\Lambda)$ and it is not if $\rho > R(\Lambda)$; obviously, $R(\mathbb{Z}) = \pi$.

The spectral radious has a definition in terms of complex analysis, by means of a duality argument. Indeed, incompleteness is equivalent to the existence of $f \in L^2([-\rho, +\rho])$, $f \neq 0$, orthogonal to $\mathcal{E}(\Lambda)$, which in terms of the Fourier–Laplace transform of f

$$F(z) = \int_{-\rho}^{+\rho} f(x)e^{-ixz}\, dx,$$

means that F vanishes on Λ, yet F is not identically zero. The space of the F is known to be the *Paley–Wiener space* PW_ρ, consisting of the entire functions F such that

$$|F(z)| = O(e^{\rho|z|}), \qquad \int_{-\infty}^{+\infty} |F(x)|^2\, dx < +\infty.$$

Hence

$$R(\Lambda) = \inf\{\rho > 0\colon\ \exists F \in PW_\rho, F \neq 0, \text{ such that } F_{|\Lambda} = 0\},$$

or equivalently

$$R(\Lambda) = \sup\{\rho > 0\colon\ \Lambda \text{ is a uniqueness set for } PW_\rho\},$$

meaning for *uniqueness sets* of a given class of functions X those sets S such that $h = 0$ whenever $h \in X$ and h vanishes on S.

It is well–known that the value of $R(\Lambda)$ is unaffected if one replaces the L^2-norm by another reasonable norm, for instance the sup–norm. Accordingly, in the complex analysis description above one may use other spaces instead of the Paley–Wiener spaces, for instance, their L^∞-versions, the Bernstein spaces consisting of entire functions F bounded in the real line and such that $|F(z)| = O(e^{\rho|z|})$.

Hence, what we have proved before is that if φ is a Λ-generator for $L^1(\mathbb{R})$ then $R(\Lambda) = +\infty$, a fact that, as mentioned before, is far from being true for $L^2(\mathbb{R})$.

The description of the spectral radious $R(\Lambda)$ in geometric or metric terms was one of the main problems in classical harmonic analysis during the first half of last century, till Beurling and Malliavin gave a solution in terms of the so–called *Beurling–Malliavin exterior density* $D_{BM}(\Lambda)$. Its definition is not simple; here it will suffice to point out that it is related to the ordinary upper density of Polya,

$$\overline{D}(\Lambda) = \limsup_{r\to+\infty} \frac{n_\Lambda(r)}{2r},$$

where $n_\Lambda(r)$ denotes the number of points of Λ in $(-r, +r)$. In fact one has $D_{BM}(\Lambda) \geq \overline{D}(\Lambda)$, although it may happen that $\overline{D}(\Lambda) = 0$ while the Beurling–Malliavin density is infinite. In a series of celebrated papers, Beurling and Malliavin proved that $R(\Lambda) = \pi D_{BM}(\Lambda)$ (see [7] for all these topics).

4. Relation with quasianalytic classes

Let us look at the dual formulation in the case of $L^1(\mathbb{R})$. By Hann–Banach theorem, φ is a Λ-generator of $L^1(\mathbb{R})$ if and only if whenever $h \in L^\infty(\mathbb{R})$ satisfies

$$\int_{-\infty}^{+\infty} h(x)\varphi(x-\lambda)\,dx = 0, \quad \lambda \in \Lambda,$$

one has $h \equiv 0$. Now notice that the above integral is the convolution $h * \check{\varphi}$, where $\check{\varphi}(x) = \varphi(-x)$. Notice also that the statement $h \equiv 0$ is equivalent to the statement $h * \check{\varphi} \equiv 0$; this is a consequence of Wiener's tauberian theorem, because the subspace E of $L^1(\mathbb{R})$ consisting of those Ψ for which $h * \Psi \equiv 0$ is translation invariant and contains $\check{\varphi}$, whose Fourier transform never vanishes; hence E is the whole $L^1(\mathbb{R})$ and therefore $h * \check{\varphi} \equiv 0$ implies $h \equiv 0$. In conclusion, we may say that φ is a Λ-generator of $L^1(\mathbb{R})$ if and only $\hat{\varphi}$ never vanishes and Λ is a uniqueness set for the class $Y = L^\infty(\mathbb{R}) * \check{\varphi}$.

If one is to search for classes of functions having discrete uniqueness sets, it is quite natural to consider classes of analytic functions, or more generally, quasianalytic classes. Among the different definitions of those we use here the *Denjoy–Carleman class* $C\{M_n\}$ associated to a sequence of positive numbers (M_n). It consists of all $f \in C^\infty(\mathbb{R})$ such that

$$|f^{(n)}(x)| \le C_f \beta_f^n M_n, \quad n = 0, 1, 2, \ldots, \quad x \in \mathbb{R}.$$

Without loss of generality the sequence (M_n) can be assumed to be log–convex (that is, $M_0 = 1$, $M_n^2 \le M_{n-1}M_{n+1}$, see [7]), implying that $M_n^{1/n}$ is increasing. This class is called *quasianalytic* if $f \in C\{M_n\}$ and $f^{(n)}(0) = 0$ $\forall n$ implies $f \equiv 0$; this is the case if and only if

$$\sum_{n=1}^{\infty} \frac{M_{n-1}}{M_n} = +\infty.$$

An equivalent condition is the divergence of the series with general term $M_n^{-1/n}$. Among the quasianalytic classes, the analytic ones (consisting of entire functions) correspond to sequences (M_n) such that $(M_n/n!)^{1/n}$ tends to zero.

Generally speaking, a quasianalytic class $C\{M_n\}$ has *discrete uniqueness sets* Λ, whose description depends on a certain density depending on the sequence (M_n) (see next section).

All this gives a way to construct concrete examples of $L^1(\mathbb{R})$-generators and spectral sets. Namely, we should choose a quasianalytic class $C\{M_n\}$ and φ with non-vanishing Fourier transform such that $L^\infty(\mathbb{R}) * \check{\varphi}$ is included in $C\{M_n\}$. Then, if Λ is a uniqueness set for $C\{M_n\}$, φ will be a Λ-generator for $L^1(\mathbb{R})$ and so Λ is spectral for $L^1(\mathbb{R})$. However, it is immediate using Cauchy inequalities that if $M_n^{1/n}$ is bounded then $C\{M_n\}$ is included in some Bernstein class B_ρ, and conversely, a Bernstein class is included in some $C\{M_n\}$ with $M_n^{1/n}$ bounded; this means that the uniqueness sets of the quasianalytic classes $C\{M_n\}$ with $M_n^{1/n}$ bounded are exactly the uniqueness sets for some Bernstein class and these, as explained before, are exactly those with $R(\Lambda) > 0$. Since we already know that $R(\Lambda) = +\infty$ is a necessary condition, this means that this approach can work only in case $M_n^{1/n} \to +\infty$. In the same direction, note that the obvious condition on φ ensuring that $L^\infty(\mathbb{R}) * \check{\varphi}$ is included in $C\{M_n\}$, namely

$$\int_{-\infty}^{+\infty} |\varphi^{(n)}(x)|\, dx \leq C M_n$$

implies $|\xi|^n |\hat{\varphi}(\xi)| \leq M_n$ and so $M_n^{1/n} \to +\infty$.

The proof along these lines that the uniqueness sets of a quasianalytic class $C\{M_n\}$ with $M_n^{1/n} \to +\infty$ are spectral sets for $L^1(\mathbb{R})$ is to be found in [5].

5. The spectral sets for $L^1(\mathbb{R})$

We have seen in Section 2 that the spectral sets of $L^1(\mathbb{R})$ have infinite spectral radious and in Section 3 we have explained why the uniqueness sets for certain analytic classes are spectral sets for $L^1(\mathbb{R})$. These three concepts turn out to be equivalent, so the following theorem can be stated:

Theorem 5.1. *For a discrete set $\Lambda \subset \mathbb{R}$, the following conditions are equivalent:*

(a) It is a spectral set for $L^1(\mathbb{R})$, that is, there exists a Λ-generator φ for $L^1(\mathbb{R})$.
(b) The spectral radius of Λ is $+\infty$.
(c) Λ is a uniqueness set for a quasianalytic class $C\{M_n\}$ with $M_n^{1/n} \nearrow \infty$.

Moreover, the generator φ can be chosen in $C\{M_n\}$.

The remaining implication (b)⇒(c), the hardest one, is proved in full generality in [4] (however, given Λ with $R(\Lambda) = +\infty$ one can construct directly a Λ-generator φ for $L^1(\mathbb{R})$ without appealing to (c), see also [4]).

The proof in [4] depends on the Beurling–Malliavin result, that is, one shows that if $D_{BM}(\Lambda) = +\infty$ then Λ is a uniqueness set for a certain quasianalytic class. This proof uses the methods in the Beurling–Malliavin theorem (see [7]), in fact one shows uniqueness for certain "generalized Bernstein" classes of analytic functions. However, under the stronger hypothesis that $\overline{D}(\Lambda) = +\infty$ one can give a simpler proof and in doing so reviewing some facts about uniqueness sets for quasianalytic classes.

Hirschman ([6]) studied them for quasianalytic non analytic classes. For *general* quasianalytic classes, a uniqueness criteria can be obtained from a very interesting lemma of Bang ([2]). Bang's lemma states that if g is C^∞ in $[-1,1]$ and $\|g^{(n)}\|_\infty \leq M_n$, with M_n log–convex, then the number of zeros of g (counting multiplicities) in $[-1,1]$ does not exceed the so–called *Bang's number* of g, which is defined as the largest integer N such that

$$\sum_{\log \|g\|_\infty^{-1} < n \leq N} \frac{M_{n-1}}{M_n} < 2e.$$

From Bang's lemma, one can obtain a result of the kind we are looking. For a given quasianalytic class $C\{M_n\}$, let $\bar{M}[k]$ denote the sequence of partial sums

$$\bar{M}[k] = \sum_{n=1}^{k} \frac{M_{n-1}}{M_n}$$

so that $\bar{M}[k] \nearrow +\infty$ as $k \nearrow +\infty$. A rescaling of Bang's lemma gives that the condition

$$\limsup_{r \nearrow \infty} \frac{\bar{M}[n_\Lambda(r)]}{2r} > e$$

implies that Λ is a uniqueness set for $C\{M_n\}$. Now it is enough to notice that given Λ with $\overline{D}(\Lambda) = +\infty$ one can construct a sequence (M_n) such that the above density is infinite.

6. The generators for $L^1(\mathbb{R})$

Of course, a very natural question is to ask for a characterization of the generators for $L^1(\mathbb{R})$, that is, which condition(s) besides the non-vanishing of $\hat{\varphi}$ describe them. This seems to be a difficult question that we now comment.

It is not difficult however to state conditions ensuring that φ is not a generator, even with the obvious necessary condition $\hat{\varphi}(\xi) \neq 0$ for all ξ. For instance, this is trivially the case if φ is compactly supported, for then just a finite number of the $\{\tau_\lambda \varphi \colon \lambda \in \Lambda\}$ will not be identically zero in a fixed interval. A non trivial result, due to Ulanovski, can be obtained from the observation that φ will not be a generator if for each $\varepsilon > 0$ there exists $f \in L^\infty(\mathbb{R})$, $f \not\equiv 0$, such that $f * \check{\varphi}$ is supported in $(-\varepsilon, \varepsilon)$, because this forces every possible uniqueness set for $L^\infty(\mathbb{R}) * \check{\varphi}$ to be everywhere dense. The so–called Beurling–Malliavin multiplier theory can be used for this purpose. Recall that a weight $\omega \leq 1$ on $\mathbb{R}$ is said to *admit multipliers* ([7]) if there exist entire functions G_ε of arbitrarily small exponential type ε for which $\omega(x)G_\varepsilon(x)$ is bounded or belongs to $L^p(\mathbb{R})$. Then it can be proved (see [5]) that φ is not a generator if $|\hat{\varphi}|^{-1}$ admits multipliers. A necessary condition on a weight ω to admit multipliers is the convergence of the logarithmic integral

$$\int_{-\infty}^{+\infty} \frac{\log(\omega(x))}{1+x^2}\, dx < \infty.$$

This condition is sufficient when ω satisfies some regularity assumptions, in particular when ω is even and increasing (see [7] for all these facts). So, if $|\hat{\varphi}|$ is even, decreasing along the positive axis and the logarithmic integral converges, φ is not a generator.

On the other hand, all generators exhibited in Section 3 have infinite logarithmic integral. This is because $|\xi|^n |\hat{\varphi}(\xi)| \leq M_n$ implies

$$|\hat{\varphi}(\xi)| \leq \inf_n \frac{M_n}{|\xi|^n} \stackrel{\text{def}}{=} M(\xi)$$

and

$$\int_{-\infty}^{+\infty} \frac{\log M(\xi)}{1+\xi^2}\, d\xi = -\infty$$

is equivalent to the quasianalyticity condition of the class.

The divergence of the logarithmic integral,

$$\int_{-\infty}^{+\infty} \frac{\log |\hat{\varphi}(\xi)|}{1+\xi^2}\, d\xi = -\infty$$

is essentially necessary for *quasianalytic* generators, that is, such that $L^\infty(\mathbb{R}) * \check{\varphi} \subset C\{M_n\}$ with $C\{M_n\}$ quasianalytic.

So all results known up to now seem to indicate that

$$\hat{\varphi}(\xi) \neq 0, \quad \int_{-\infty}^{+\infty} \frac{\log |\hat{\varphi}(\xi)|}{1+\xi^2}\, d\xi = -\infty$$

might be a characterization of the generators for $L^1(\mathbb{R})$.

7. The restricted problem

Another interesting problem arises when we look at a specific generator φ for $L^1(\mathbb{R})$ and ask for which discrete sets Λ the family $\Lambda(\varphi)$ spans $L^1(\mathbb{R})$. According to what has been said in Section 3, these sets are exactly the uniqueness sets of the function space $Y_\varphi = L^\infty(\mathbb{R}) * \check{\varphi}$, so an exact knowledge of Y_φ seems adequate.

One case in which the space Y_φ can be exactly described is the one of the Poisson function

$$\varphi(t) = \frac{1}{\pi}\frac{1}{1+t^2}.$$

This is achieved in [3], and we proceed to explain the main ideas. We are to describe the functions of type $F = f * \varphi$, $f \in L^\infty(\mathbb{R})$ and real. In order to do that we notice first that F makes sense for complex z with $|\text{Im } z| < 1$,

$$F(z) = \frac{1}{\pi}\int_{-\infty}^{+\infty} \frac{1}{1+(t-z)^2}\, dt, \quad f \in L^\infty(\mathbb{R})$$

and that this defines an holomorphic function in the strip $B = \{z \colon |\text{Im } z| < 1\}$ with bounded real part, and real in the real line, that is, $F(\overline{z}) = \overline{F(z)}$. We call $E^\infty(B)$ the class of such F. In [3], it is proved that, conversely, any function $F \in E^\infty(B)$ can be expressed as above with f a bounded function.

Hence the problem for the Poisson function becomes the problem of describing the uniqueness real sets for the class $E^\infty(B)$. By transferring the problem to the unit disk D by means of a suitable conformal map, it is not hard to check that the required condition is nothing else that the usual Blashke condition describing the uniqueness sets $(\mu_n)_n$ for the Hardy classes $H^p(D)$, namely $\sum_n (1 - |\mu_n|) = +\infty$. When transferring again to B this condition becomes

$$\sum_{n\in\mathbb{Z}} e^{-\frac{\pi}{2}|\lambda_n|} = +\infty,$$

so this is the exact description of the discrete sets $\Lambda = \{\lambda_n \colon n \in \mathbb{Z}\}$ such that $\Lambda(\varphi)$ spans $L^1(\mathbb{R})$.

Strictly speaking one does not need to know exactly the function space $Y_\varphi = L^\infty(\mathbb{R}) * \check{\varphi}$ to describe its uniqueness sets, as often it is intermediate between two spaces Y_1, Y_2 having the same uniqueness sets. This technique can be used to deal with generators φ of Poisson type.

Another case which is worth mentioning is the Gaussian function $\varphi(x) = e^{-x^2}$. When dealing with the problem in $L^2(\mathbb{R})$, the corresponding space

$Y = L^2(\mathbb{R}) * \varphi$ can be identified exactly with the Fock space of entire functions F for which

$$\int_{\mathbb{C}} |F(z)|^2 e^{-|z|^2}\, dm(z) < +\infty.$$

Replacing $L^2(\mathbb{R})$ by $L^\infty(\mathbb{R})$ leads to an L^∞ version of the Fock space. In any event, the description of the uniqueness sets for the Fock space is an unsolved problem; this description is probably not possible just with size conditions, as subtle equilibrium conditions play into role. The best size condition on $\Lambda = \{\lambda_n : n \in \mathbb{Z}\}$ ensuring that it is a uniqueness set, and hence $\Lambda(\varphi)$ spans $L^1(\mathbb{R})$, is due to [14] and reads

$$\sum_n \frac{1}{|\lambda_n|^2} = +\infty.$$

References

1. A. Atzmon and A. Olevskii, *Completeness of integer translates in function spaces on* $\mathbb{R}$, J. Approximation Theory **83**(3) (1996), 291–327.
2. T.Bang, *The theory of metric spaces applied to infinitely differentiable functions*, Math. Scand. **1** (1953), 137–152.
3. J. Bruna and M. Melnikov, *On translates of the Poisson kernel and zeros of harmonic functions*, to appear in the Bull. London Math. Soc.
4. J. Bruna, A. Olevskii and A. Ulanovskii, *Completeness in* $L^1(\mathbb{R})$ *of discrete translates*, Revista Matemática IberoAmericana **22**(1) (2006), 1–16.
5. J. Bruna, *On translation and affine systems spanning* $L^1(\mathbb{R})$, J. of Fourier Anal. Appl. **12**(1) (2006), 71–82.
6. I. I. Hirschman, *On the distribution of zeros of functions belonging to certain quasianalytic classes*, Amer. J. Math. **72** (1950), 396–406.
7. P. Koosis, *The Logarithmic Integral, Vol. I-II*, Cambridge Univ. Press (1992).
8. H. J. Landau, *A sparse sequence of exponentials closed on large sets*, Bull. Amer. Math. Soc. **70** (1964), 566–569.
9. N. Nikolskii, *Selected problems of weighted approximation and spectral analysis*, Trudy Math. Inst. Steklov **120** (1974). English translation in Proc. Steklov Math. Inst. **120** (1974).
10. H. J. Landau, *A sparse sequence of exponentials closed on large sets*, Bull. Amer. Math. Soc. **70** (1964), 566–569.
11. A. Olevskii, *Completeness in* $L^2(\mathbb{R})$ *of almost integer translates*, C. R. Acad. Sci. Paris **324**(1) (1997), 987–991.
12. A. Olevskii and A. Ulanovskii, *Almost integer translates, do nice generators exists?*, J. Fourier Anal. Appl. **10**(1) (2004), 93–104.
13. A. Ulanovskii, *Sparse systems of functions closed on large sets in* $\mathbb{R}^n$, J. London Math. Soc **63**(2) (2001), 428–440.
14. R. A. Zalik, *On approximation by shifts and a theorem of Wiener*, Trans. Amer. Math. Soc. **243** (1978), 299–308.

A NEW LOOK AT COMPACTNESS VIA DISTANCES TO FUNCTION SPACES

CARLOS ANGOSTO[†] and BERNARDO CASCALES[‡]

Department of Mathematics
Universidad de Murcia
30100 Espinardo (Murcia), Spain
E-mails: [†] *angosto@um.es*, [‡] *beca@um.es*

Many classical results about compactness in functional analysis can be derived from suitable inequalities involving distances to spaces of continuous or Baire one functions: this approach gives an extra insight to the classical results as well as triggers a number of open questions in different exciting research branches. We exhibit here, for instance, *quantitative* versions of Grothendieck's characterization of weak compactness in spaces $C(K)$ and also of the Eberlein–Šmulyan and Krein–Šmulyan theorems. The above results specialized in Banach spaces lead to several equivalent measures of *non-weak compactness*. In a different direction we envisage a method to measure the distance from a function $f \in \mathbb{R}^X$ to $B_1(X)$ — space of Baire one functions on X — which allows us to obtain, when X is Polish, a quantitative version of the well-known Rosenthal's result stating that in $B_1(X)$ the pointwise relatively countably compact sets are pointwise compact. Other results and applications are commented too.

Keywords: Eberlein–Grothendieck theorem, Krein–Smulyan theorem, oscillations, iterated limits, compactness, measures of non compactness, distances to function spaces, Rosenthal theorem, Baire one functions.

1. Introduction

These are the written notes of a lecture with the same title delivered by the second named author at the *III International Course of Mathematical Analysis of Andalucía, Huelva, September 3-7, 2007.* We collect here results, mostly without proof, that mainly correspond to the papers [3–5,11]. A good deal of extra information about the subject can also be found in the Ph.D. dissertation by the first named author ([1]).

In this *survey*, we present recent *quantitative* versions of many of the classical compactness results in functional analysis and their relatives. As an example and in order to fix ideas, one of the problems studied is illustrated and explained in the lines below. Take K a compact Hausdorff space and let $C(K)$ be the space of real–valued continuous functions defined on K. Look at $C(K)$ embedded in $\mathbb{R}^K$, let d be the metric of uniform convergence on $\mathbb{R}^K$ and take $H \subset \mathbb{R}^K$ a uniformly bounded set. If τ_p is the topology of pointwise convergence on $\mathbb{R}^K$, then Tychonoff's theorem says that $\overline{H}^{\mathbb{R}^K}$ is τ_p-compact. Therefore, for H to be τ_p-relatively compact in $C(K)$, the only thing we should worry about is to have $\overline{H}^{\mathbb{R}^K} \subset C(K)$. Notice that if $\hat{d}$ is the *worst* distance from $\overline{H}^{\mathbb{R}^K}$ to $C(K)$ then $\hat{d} = 0$ if and only if $\overline{H}^{\mathbb{R}^K} \subset C(K)$. In general $\hat{d} \geq 0$ gives us a measure of non τ_p-compactness for H relative to $C(K)$. Hence the questions are: *a) Is there any way of computing $\hat{d}$?;*

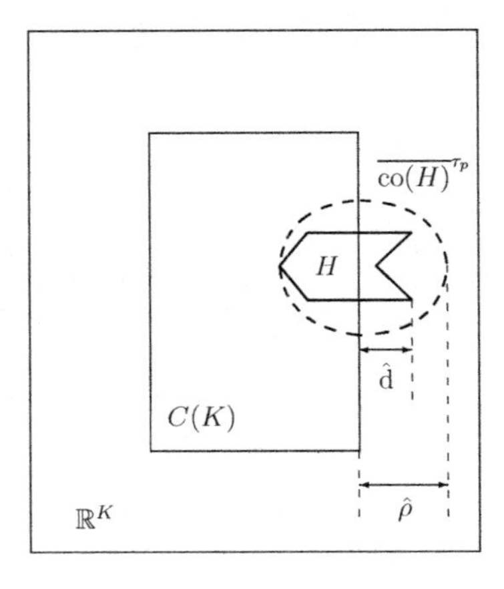

Fig. 1.

b) are there useful estimates involving $\hat{d}$ that are equivalent to qualitative properties of the sets H's? The answer to *a)* has been known for a long time and is yes: the distance of a function $f \in \mathbb{R}^K$ to $C(K)$ can be computed in terms of the global oscillation of f on K, see Section 2. Here is a first case in the spirit of *b)* that is illustrated through the Figure 1: if $\hat{\rho}$ is the *worst* distance from the closed convex hull $\overline{\mathrm{co}\,(H)}^{\mathbb{R}^K}$ to $C(K)$, then it is proved that $\hat{d} \leq \hat{\rho} \leq 5\hat{d}$ (the constant 5 can be replaced by 2 for sets $H \subset C(K)$). Note that the above inequality is the quantitative version of the celebrated Krein–Šmulyan theorem about weak compactness of the closed convex hull of weakly compact sets in Banach spaces.

A bit of the history behind the classical results that we quantify follows. In 1940 Šmulyan ([30]) showed that weakly relatively compact subsets of a Banach space are weakly relatively sequentially compact. He also proved that if a Banach space E has w^*-separable dual then a subset H of E is weakly relatively countably compact if and only if H is weakly relatively sequentially compact. Dieudonné and Schwartz ([14]) extended this last result to locally convex spaces with a coarser metrizable topology. The converse of Šmulyan theorem was obtained by Eberlein ([15]) who proved that relatively countably compact sets are relatively compact sets for the weak topology of a Banach space. Grothendieck generalized these results to locally convex spaces that are quasicomplete for its Mackey topology: this result is based upon a similar one for spaces $(C(K), \tau_p)$ of continuous functions on a compact space K endowed with the pointwise convergence topology. Fremlin's notion of angelic space and some of its consequences can be used for proving the above results in a clever and clear way, see the book by Floret [17]. Orihuela ([27]) showed in 1987 that spaces $(C(X), \tau_p)$ with X a countably K-determined space (or even more general spaces) are angelic. Similarly, for spaces $(B_1(X), \tau_p)$ of Baire one functions on a Polish space with the pointwise convergence topology, Rosenthal showed that relatively countably compact sets are relatively compact. Bourgain, Fremlin and Talagrand ([10]) showed that, in fact, $(B_1(X), \tau_p)$ is angelic.

In recent years, several *quantitative* counterparts for some other classical results have been proved by different authors. These new versions strengthen the original theorems and lead to new problems and applications in topology and analysis: see, for instance, [16,18–21].

A bit of terminology: by letters $T, X, Y, \dots$ we denote sets or completely regular topological spaces; (Z, d) is a metric space (Z if d is tacitly assumed); $\mathbb{R}$ is considered as a metric space endowed with the metric associated to the absolute value $|\cdot|$. The space Z^X is equipped with the product topology τ_p. We let $C(X, Z)$ denote the space of all Z-valued continuous functions on X, and let $B_1(X, Z)$ denote the space of all Z-valued functions of the first Baire class (Baire one functions), i.e., pointwise limits of Z-valued continuous functions. When $Z = \mathbb{R}$, we write, as usual, $C(X)$ and $B_1(X)$ for $C(X, \mathbb{R})$ and $B_1(X, \mathbb{R})$, respectively.

If $\emptyset \neq A \subset (Z, d)$ we write diam $(A) := \sup\{d(x, y)\colon x, y \in A\}$. For A and B nonempty subsets of (Z, d), we consider the *usual distance* between A and B given by

$$d(A, B) = \inf\{d(a, b)\colon a \in A, b \in B\},$$

and the *Hausdorff non-symmetrized distance* from A to B defined by

$$\hat{d}(A, B) = \sup\{d(a, B)\colon a \in A\}.$$

In Z^X we deal with the *standard supremum metric* given for arbitrary functions $f, g \in Z^X$ by

$$d(f,g) = \sup_{x \in X} d(f(x), g(x))$$

that is allowed to take the value $+\infty$. If $\mathcal{F} \subset Z^X$ is some space of functions we consequently define $d(f, \mathcal{F})$ and $\hat{d}(A, \mathcal{F})$ for sets $A \subset Z^X$; the spaces of functions $\mathcal{F}$ that we will consider are $C(X, Z)$ and $B_1(X, Z)$.

By $(E, \|\cdot\|)$ we denote a real Banach space (or simply E if $\|\cdot\|$ is tacitly assumed). Finally, B_E stands for the closed unit ball in E, E^* for the dual space of E and E^{**} for the bidual space of E; w is the weak topology of a Banach space and w^* is the weak* topology in the dual.

2. Distance to spaces of continuous functions

We start with the proof for the formula (1) below that gives us the distance of a function $f \in \mathbb{R}^X$ to the space of continuous functions $C(X)$. Next result is used in the proof that we provide for Theorem 2.2.

Theorem 2.1 ([23, Theorem 12.16]). *Let X be a normal space and let $f_1 \leq f_2$ be two real functions on X such that f_1 is upper semicontinuous and f_2 is lower semicontinuous. Then, there exists a continuous function $f \in C(X)$ such that $f_1(x) \leq f(x) \leq f_2(x)$ for all $x \in X$.*

Theorem 2.2. *Let X be a normal space. If $f \in \mathbb{R}^X$, then*

$$d(f, C(X)) = \frac{1}{2}\text{osc }(f) \tag{1}$$

where

$$\text{osc }(f) = \sup_{x \in X} \text{osc }(f,x) = \sup_{x \in X} \inf\{\text{diam } f(U) \colon U \subset X \ \ \textit{open}, x \in U\}.$$

Proof. We prove first that $\frac{1}{2}\text{osc }(f) \leq d(f, C(X))$. If $d(f, C(X))$ is infinite, the inequality clearly holds. Suppose that $\rho = d(f, C(X))$ is finite. Fix $x \in X$ and $\varepsilon > 0$. Take $g \in C(X)$ such that $d(f, g) \leq \rho + \varepsilon/3$. Since g is continuous at x, there is an open neighborhood U of x such that diam $(g(U)) < \varepsilon/3$. Then, if $y, z \in U$,

$$d(f(y), f(z)) \leq d(f(y), g(y)) + d(g(y), g(z)) + d(g(z), f(z)) < 2\rho + \varepsilon.$$

Thus osc $(f, x) < 2\rho + \varepsilon$ for each $\varepsilon > 0$. We conclude that osc $(f, x) \leq 2\rho$ for every $x \in X$ and so, the inequality $\frac{1}{2}\text{osc }(f) \leq d(f, C(X))$ is established.

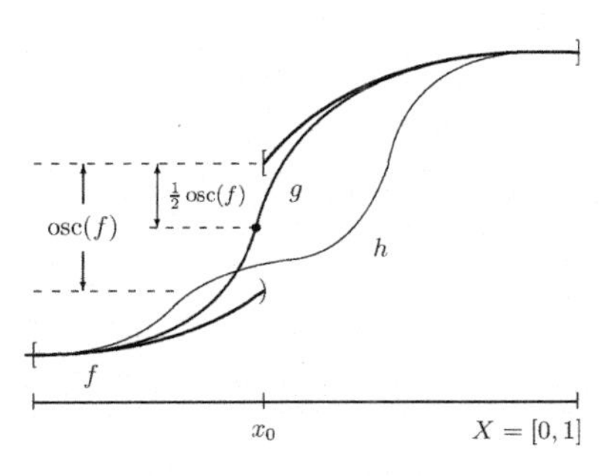

Fig. 2.

Let us prove now that $d(f, C(X)) \leq \frac{1}{2}\text{osc}\ (f)$. We only have to prove the inequality when $\delta = \frac{1}{2}\text{osc}\ (f)$ is finite. For $x \in X$, denote by $\mathcal{U}_x$ the family of open neighborhoods of x and define

$$\mathcal{V}_x := \{U \in \mathcal{U}_x : \text{diam}\ (f(U)) < \text{osc}\ (f) + 1\}.$$

Clearly $\mathcal{V}_x$ is a basis of neighborhoods for x and for each $U \in \mathcal{V}_x$, $f|_U$ is upper and lower bounded.

An easy computation gives us that

$$\begin{aligned} 2\delta \geq \text{osc}\ (f, x) &= \inf_{U \in \mathcal{U}_x} \text{diam}\ (f(U)) = \inf_{U \in \mathcal{V}_x} \text{diam}\ (f(U)) \\ &= \inf_{U \in \mathcal{V}_x} \sup_{y,z \in U} (f(y) - f(z)) \\ &\geq \inf_{U,V \in \mathcal{V}_x} \sup_{y \in U, z \in V} (f(y) - f(z)) \\ &= \inf_{U \in \mathcal{V}_x} \sup_{y \in U} f(y) - \sup_{U \in \mathcal{V}_x} \inf_{z \in U} f(z). \end{aligned}$$

If we define

$$f_1(x) := \inf_{U \in \mathcal{V}_x} \sup_{z \in U} f(z) - \delta$$

$$f_2(x) := \sup_{U \in \mathcal{V}_x} \inf_{z \in U} f(z) + \delta$$

then $f_1 \leq f_2$. It is easy to check that f_1 is upper semicontinuous and f_2 is lower semicontinuous. By Theorem 2.1, there is a continuous function $h \in C(X)$ such that

$$f_1(x) \leq h(x) \leq f_2(x)$$

for every $x \in X$. On the other hand, for every $x \in X$ we have

$$f_2(x) - \delta \leq f(x) \leq f_1(x) + \delta$$

and therefore

$$h(x) - \delta \leq f_2(x) - \delta \leq f(x) \leq f_1(x) + \delta \leq h(x) + \delta.$$

So $d(f,h) \leq \delta = \frac{1}{2}\text{osc}\,(f)$ and this finishes the proof. $\square$

A proof for the above result when X is a paracompact space and all functions are assumed to be bounded can be found in [9, Proposition 1.18]. We note that the validity of Theorem 2.2 characterizes normal spaces.

Corollary 2.1. *Let X be a topological space. The following statements are equivalent:*

(i) X is normal,
(ii) for each $f \in \mathbb{R}^X$ there is $g \in C(X)$ such that $d(f,g) = \frac{1}{2}\text{osc}\,(f)$,
(iii) $d(f, C(X)) = \frac{1}{2}\text{osc}\,(f)$ for each function $f \in \mathbb{R}^X$.

3. Distances to spaces of continuous functions on compact spaces

We aim now to estimate $\hat{d} = \hat{d}(\overline{H}^{\mathbb{R}^K}, C(K))$ using some other distinguished *quantities* that we shall define.

Let T be a topological space. For a subset A of T, $A^{\mathbb{N}}$ is considered as the set of all sequences in A and the set of all cluster points in T of a sequence $\varphi \in A^{\mathbb{N}}$ is denoted by $\text{clust}_T(\varphi)$. Recall that $\text{clust}_T(\varphi)$ is a closed subset of T that can be expressed as

$$\text{clust}_T(\varphi) = \bigcap_{n\in\mathbb{N}} \overline{\{\varphi(m) \colon m > n\}}.$$

Definition 3.1. Let X be a topological space and (Z,d) a metric space. If H is a subset Z^X we define

$$\text{ck}\,(H) := \sup_{\varphi\in H^{\mathbb{N}}} d(\text{clust}_{\mathbb{R}^K}(\varphi), C(X,Z)).$$

If $K \subset X$ we write

$$\gamma_K(H) := \sup\left\{d(\lim_n \lim_m f_m(x_n), \lim_m \lim_n f_m(x_n)) \colon (f_m) \subset H, (x_n) \subset K\right\},$$

assuming the involved limits exist.

By definition, we agree that $\inf \emptyset = +\infty$. Observe that if $H \subset C(X,Z)$ is a τ_p-relatively countably compact subset of $C(X,Z)$ then $\text{ck}\,(H) = 0$. Also notice that $\gamma_K(H) = 0$ means in the language of [22] that H interchanges limits with K.

Theorem 3.1 ([3,11]). *Let K be a compact space and let H be a uniformly bounded subset of $C(K)$. We have*

$$\operatorname{ck}(H) \overset{(a)}{\leq} \hat{d}(\overline{H}^{\mathbb{R}^K}, C(K)) \overset{(b)}{\leq} \gamma_K(H) \overset{(c)}{\leq} 2\operatorname{ck}(H).$$

Explanation of the proof. The details of the proof can be found in [3,11]. Here is a pretty short explanation of the ideas behind. Inequality (a) straightforwardly follows from the definitions involved. Inequality (c) uses the same kind of arguments than those used in the proof to show that if H is τ_p-relatively compact in $C(K)$ then H interchanges limits with K. Inequality (b) is much more involved than the other two: here the idea is to show that for every $x \in K$ and $f \in \overline{H}^{\mathbb{R}^K}$, the semioscillation

$$\operatorname{osc}^*(f,x) := \inf_U \{\sup_{y \in U} |f(y) - f(x)| : U \subset X \text{ open}, x \in U\}$$

is at most $\gamma_K(H)$. Therefore, $\operatorname{osc}(f) \leq 2\gamma_K(H)$ and now Theorem 2.2 applies to finally obtain that $d(f, C(K)) \leq \gamma_K(H)$. Thus $d(\overline{H}^{\mathbb{R}^K}, C(K)) \leq \gamma_K(H)$ and (a) is proved. □

The following theorem is a quantitative version of the Krein–Šmulyan theorem: see next section for its consequences in Banach spaces.

Theorem 3.2 ([11]). *Let K be a compact topological space and let H be a uniformly bounded subset of $\mathbb{R}^K$. Then*

$$\gamma_K(H) = \gamma_K(\operatorname{co}(H)) \tag{2}$$

and, as a consequence, for $H \subset C(K)$ we obtain that

$$\hat{d}(\overline{\operatorname{co}(H)}^{\mathbb{R}^K}, C(K)) \leq 2\hat{d}(\overline{H}^{\mathbb{R}^K}, C(K)) \tag{3}$$

and if $H \subset \mathbb{R}^K$ is uniformly bounded then

$$\hat{d}(\overline{\operatorname{co}(H)}^{\mathbb{R}^K}, C(K)) \leq 5\hat{d}(\overline{H}^{\mathbb{R}^K}, C(K)). \tag{4}$$

Explanation of the proof. The equality (2) is rather involved: the proof offered in [11] uses some ideas from the proof of the Krein–Smulyan theorem in Kelley–Namioka's book [25, Chapter 5, Section 17]; we note that a version for Banach spaces, less general than the one here, was proved first in [16] using Ptak's combinatorial lemma. Inequality (3) easily follows from (2) and Theorem 3.1:

$$\hat{d}(\overline{\operatorname{co}(H)}^{\mathbb{R}^K}, C(K)) \leq \gamma_K(\operatorname{co}(H)) = \gamma_K(H) \leq 2\operatorname{ck}(H) \leq 2\hat{d}(\overline{H}^{\mathbb{R}^K}, C(K)).$$

When $H \subset \mathbb{R}^K$, we approximate H by some set in $C(K)$, then use inequality (3) and, after some computations with the sets, *5* appears as $5 = 2\times 2+1$: see [11] for details. □

4. Distance to Banach spaces

The aim of this section is to specialize the result of the previous one in the case of Banach spaces: in order to do so, we have to overcome some technicalities. If E is Banach space and H is a bounded subset of E and we consider the w^*-closure of H in E^{**}, we can measure how far H is from being w-relatively compact in E using

$$k(H) = \hat{d}(\overline{H}^{w^*}, E) = \sup_{y\in\overline{H}^{w^*}} \inf_{x\in E} \|y - x\|.$$

Next theorem gives as a tool to export results obtained in the context of distances to spaces of continuous functions on a compact set to the context of Banach spaces.

Theorem 4.1 ([11]). *Let E be a Banach space and let B_{E^*} be the closed unit ball in the dual E^* endowed with the w^*-topology. Let $i\colon E \to E^{**}$ and $j\colon E^{**} \to \ell_\infty(B_{E^*})$ be the canonical embeddings. Then, for every $x^{**} \in E^{**}$ we have*

$$d(x^{**}, i(E)) = d(j(x^{**}), C(B_{E^*})).$$

Explanation of the proof. The proof of this result goes along the proof we have given for Theorem 2.2 but instead of using Theorem 2.1 as a tool now the concourse of Hahn–Banach theorem is required: namely, it is used Theorem 21.20 in [12], that states that if $f_1 < f_2$ are two real–valued functions defined on B_{E^*} with f_1 concave and w^*-upper semicontinuous and f_2 convex and w^*-lower semicontinuous then there exist a w^*-continuous affine function h defined on B_{E^*} such that

$$f_1(x) < h(x) < f_2(x)$$

for every $x \in B_{E^*}$. See [11] for details. □

If we consider $\ell_\infty(B_{E^*})$ as a subspace of $(\mathbb{R}^{B_{E^*}}, \tau_p)$, then the natural embedding $j\colon (E^{**}, w^*) \to (\ell_\infty(B_{E^*}), \tau_p)$ is continuous. For a bounded set $H \subset E^{**}$, the closure $\overline{H}^{w^*}$ is w^*-compact and therefore, the continuity of

j gives us that $\overline{j(H)}^{\tau_p} = j(\overline{H}^{w^*})$. So

$$\begin{aligned}\hat{d}(\overline{j(H)}^{\tau_p}, C(B_{E^*}, w^*)) &= \hat{d}(j(\overline{H}^{w^*}), C(B_{E^*}, w^*)) \\ &= \sup_{z\in\overline{H}^{w^*}} d(j(z), C(B_{E^*}, w^*)) \\ &= \sup_{z\in\overline{H}^{w^*}} d(z, i(E)) = \hat{d}(\overline{H}^{w^*}, i(E)). \qquad (5)\end{aligned}$$

Similarly we have

$$d(\overline{j(H)}^{\tau_p}, C(B_{E^*}, w^*)) = d(\overline{H}^{w^*}, i(E)). \qquad (6)$$

Definition 4.1. Let E be a Banach space and let H be a subset of E. We define:

$$\gamma(H) := \sup\{|\lim_n \lim_m f_m(x_n) - \lim_m \lim_n f_m(x_n)| \colon (f_m)_m \subset B_{E^*}, (x_n)_n \subset H\},$$

assuming the involved limits exists,

$$\mathrm{ck}\,(H) := \sup_{\varphi\in H^{\mathbb{N}}} d(\mathrm{clust}_{E^{**},w^*}(\varphi), E)$$

and

$$\omega(H) := \inf\{\varepsilon > 0 \colon H \subset K_\varepsilon + \varepsilon B_E \text{ and } K_\varepsilon \subset X \text{ is } w\text{-compact}\}.$$

The function ω was introduced by de Blasi ([13]) as a measure of weak noncompactness that can be regarded as the counterpart for the weak topology of the classical Hausdorff measure of norm noncompactness. The function γ already appeared in [7] and in [26] with an a priori different definition: in the latter, the sup is taken over all the sequences in the convex hull co (H) instead of sequences only in H; but, by Theorem 3.2, $\gamma(H) = \gamma(\mathrm{co}\,(H))$, which says that our definition for γ is equivalent to the one given in [26]. The index k has been used in [11,16,18]. Whereas ω and γ are measures of weak noncompactness in the sense of the axiomatic definition given in [8] the function k fails to satisfy $\mathrm{k}\,(\mathrm{co}\,(H)) = \mathrm{k}\,(H)$, that is one of the properties required in order to be a measure of weak noncompactness in the sense of [8]: see [18,19] for counterexamples. Nonetheless, k as well as γ and ω does satisfy the condition $\mathrm{k}\,(H) = 0$ if and only if H is relatively weakly compact in E.

All the above quantities are related with each other.

Theorem 4.2 ([4,11]). *Let H be a bounded subset of a Banach space E. Then*

$$\mathrm{ck}\,(H) \le \mathrm{k}\,(H) \le \gamma(H) \le 2\mathrm{ck}\,(H) \le 2\mathrm{k}\,(H) \le 2\omega(H) \qquad (7)$$

$$\gamma(H) = \gamma(\text{co}\,(H)) \quad \textit{and} \quad \omega(H) = \omega(\text{co}\,(H).$$

For any $x^{**} \in \overline{H}^{w^*}$, *there is a sequence* $(x_n)_n$ *in* H *such that*

$$\|x^{**} - y^{**}\| \leq \gamma(H)$$

for any cluster point y^{**} *of* $(x_n)_n$ *in* E^{**}. *Furthermore,* H *is relatively compact in* (E, w) *if and only if it is zero one (equivalently all) of the numbers* $\text{ck}\,(H)$, $\text{k}\,(H)$, $\gamma(H)$ *and* $\omega(H)$.

Explanation of the proof. The first part of the Theorem uses the results stated in the previous section together with the equalities (5) and (6). For the second part, the approximation by sequences, again equalities (5) and (6) are used together now with [11, Proposition 5.2]. □

We point out that $\gamma(H) = \gamma(\text{co}\,(H))$ and $\text{k}\,(H) \leq \gamma(H) \leq 2\text{k}\,(H)$ have also been established in [16]: note that inequalities (7) immediately imply Krein–Smulyan theorem for Banach spaces that states that the closed convex hull of a weakly compact set is again weakly compact.

Recall that a topological space T is said to be *angelic* if, whenever H is a relatively countably compact subset of T, its closure $\overline{H}$ is compact and each element of $\overline{H}$ is limit of a sequence in H: a good reference for angelic spaces is [17]. Inequalities (7) together with the approximation by sequences in Theorem 4.2 offer us a quantitative version of the angelicity of a Banach space endowed with its weak topology, Eberlein–Smulyan's theorem.

Corollary 4.1. *If* E *is a Banach space then* (E, w) *is angelic.*

In (7), the constants involved are sharp but sometimes the inequalities involved are equalities.

Theorem 4.3 ([4]). *If* E *is a Banach space with Corson property* $\mathfrak{C}$ *then for every bounded set* $H \subset E$ *we have* $\text{ck}\,(H) = \text{k}\,(H)$.

Recall that a Banach space E is said to have the Corson property $\mathfrak{C}$ if each collection of closed convex subsets of E with empty intersection has a countable subcollection with empty intersection: the class of Banach spaces with property $\mathfrak{C}$ is a wide class that contains the classes of Banach spaces which are Lindelöf for their weak topologies (in particular, w-K-analytic Banach spaces) and also the class of Banach spaces with w^*-countably tight (in particular Banach spaces with w^*-angelic dual unit ball), see [28]. We note that equality $\text{ck}\,(H) = \text{k}\,(H)$ does not hold for general Banach spaces: see [4] for a counterexample.

The Hausdorff measure of norm noncompactness is defined for bounded sets H of Banach spaces E as

$$\mathrm{h}\,(H) := \inf\{\varepsilon > 0\colon H \subset K_\varepsilon + \varepsilon B_E \text{ and } K_\varepsilon \subset X \text{ is finite}\}.$$

A theorem of Schauder states that a continuous linear operator $T\colon E \to F$ is compact if and only if its adjoint operator $T^*\colon F^* \to E^*$ is compact. A quantitative strengthening of Schauder's result was proved by Goldenstein and Marcus (cf. [7, p. 367]) who established the inequalities

$$\frac{1}{2}\mathrm{h}\,(T(B_E)) \leq \mathrm{h}\,(T^*(B_{F^*})) \leq 2\mathrm{h}\,(T(B_E)). \tag{8}$$

For weak topologies, Gantmacher established that the operator T is weakly compact if and only if T^* is weakly compact. Nonetheless, the corresponding quantitative version to (8) where h is replaced by ω fails for general Banach spaces: Astala and Tylli constructed in [7, Theorem 4] a separable Banach space E and a sequence $(T_n)_n$ of operators $T_n\colon E \to c_0$ such that

$$\omega(T_n^*(B_{\ell^1})) = 1 \qquad \text{and} \qquad \omega(T_n^{**}(B_E^{**})) \leq \omega(T_n(B_E)) \leq \frac{1}{n}.$$

On the positive side, there exists a quantitative version of Gantmacher result for γ and henceforth for k and ck.

Theorem 4.4 ([4]). *Let E and F be Banach spaces, $T\colon E \to F$ an operator and $T^*\colon F^* \to E^*$ its adjoint. Then*

$$\gamma(T(B_E)) \leq \gamma(T^*(B_{F^*})) \leq 2\gamma(T(B_E)).$$

As a combination of the result and the aforementioned Astala and Tylli's construction we obtain:

Corollary 4.2 ([4,7]). *The measures of weak noncompactness γ and ω are not equivalent, meaning, there is no $N > 0$ such that for any Banach space and any bounded set $H \subset E$ we have $\omega(H) \leq N\gamma(H)$.*

The following result is a quantitative strengthening of the classical Grothendieck's characterization of weakly compact sets in spaces $C(K)$.

Theorem 4.5 ([4]). *Let K be a compact space and let H be a uniformly bounded subset of $C(K)$. Then we have*

$$\gamma_K(H) \leq \gamma(H) \leq 2\gamma_K(H).$$

Note that this result implies that such an H is uniformly bounded subset of $C(K)$, then H is relatively weakly compact (i.e., $\gamma(H) = 0$) if and only if H is relatively τ_p-compact (i.e. $\gamma(H) = 0$). It is worth mentioning that the proof we provided in [4] does not use the Lebesgue Convergence theorem as the classical proof of Grothendieck's theorem does: our proof relies on purely topological arguments.

5. Distances to continuous functions on countably K-determined spaces

For people just interested about results for spaces of continuous functions in non compact spaces X, it is possible to get rid of the constraints imposed in Theorem 3.1 and also deal with pointwise bounded sets $H \subset \mathbb{R}^X$ instead of uniformly bounded sets made up of continuous functions. To do so, one needs to prove first the two technical lemmas that follow.

Lemma 5.1 ([3]). *Let X be a topological space, (Z, d) a metric space and H a relatively compact subset of the space (Z^X, τ_p). Then, for every relatively countably compact subset $K \subset X$ we have*

$$\gamma_K(H) \leq 2\left(\text{ck }(H) + \hat{d}(H, C(X, Z))\right).$$

Lemma 5.2 ([3]). *Suppose that (Z, d) is a separable metric space and let X be a set. Given functions $f_1, \ldots, f_n \in Z^X$ and $D \subset X$ there is a countable subset $L \subset D$ such that for every $x \in D$*

$$\inf_{y \in L} \max_{1 \leq k \leq n} d(f_k(y), f_k(x)) = 0.$$

With the above two lemmas at hand and a long way of technical difficulties to overcome one arrives to the following two results that greatly extends Theorem 3.1.

Theorem 5.1 ([3]). *Let X be a countably K-determined space, (Z, d) a separable metric space and H a relatively compact subset of the space (Z^X, τ_p). Then, for any $f \in \overline{H}^{Z^X}$, there exists a sequence $(f_n)_n$ in H such that*

$$\sup_{x \in X} d(g(x), f(x)) \leq 2\text{ck }(H) + 2\hat{d}(H, C(X, Z)) \leq 4\text{ck }(H)$$

for any cluster point g of $(f_n)_n$ in Z^X.

Theorem 5.2 ([3]). *Let X be a countably K-determined space, (Z, d) a separable metric space and H a relatively compact subset of the space (Z^X, τ_p). Then*

$$\text{ck }(H) \leq \hat{d}(\overline{H}^{Z^X}, C(X, Z)) \leq 3\text{ck }(H) + 2\hat{d}(H, C(X, Z)) \leq 5\text{ck }(H).$$

Recall that a topological space X is said to be *countably K-determined* if there is a subspace $\Sigma \subset \mathbb{N}^{\mathbb{N}}$ and an upper semicontinuous set–valued map $T\colon \Sigma \to 2^X$ such that $T(\alpha)$ is compact for each $\alpha \in \Sigma$ and $T(\Sigma) := \bigcup\{T(\alpha)\colon \alpha \in \Sigma\} = X$. Here the set–valued map T is called *upper semicontinuous* if for each $\alpha \in \Sigma$ and for any open subset U of X such that $T(\alpha) \subset U$ there exists a neighborhood V of α with $T(V) \subset U$. A good reference for countably K-determined spaces is [6] where they appear under the name *Lindelöf Σ-spaces*: notice that this class of spaces does properly contain the class of separable metric spaces and the class of K-analytic and (so) the σ-compact spaces.

We point out that the above results imply the main result in [27].

Corollary 5.1 ([27]). *Let X be a countably K-determined space and (Z, d) a metric space. Then $C_p(X, Z)$ is an angelic space.*

Our Theorems 5.1 and 5.2 can be proved (same proofs and difficulty) in the more general setting of spaces X being *web-compact, quasi–Souslin, etc.* as studied in [27]. We also notice that this quite general results can be used to obtain some consequences in the setting of locally convex spaces.

Although there are examples showing that the constants are truly needed in the inequalities in Theorem 5.2, there are cases for which k $=$ ck .

Lemma 5.3. *Let X be a first countable space, (Z, d) a metric space and H a pointwise relatively compact subset of (Z^X, τ_p). Then*

$$\sup_{f \in \overline{H}} \operatorname{osc}(f) = \sup_{\varphi \in H^{\mathbb{N}}} \inf\{\operatorname{osc}(f)\colon f \in \operatorname{clust}_{Z^X}(\varphi)\}. \tag{9}$$

For $Z = \mathbb{R}$ the equality (9) holds when X is countably tight.

The above lemma can be read as:

Proposition 5.1. *Let X be a metric space, E a Banach space and H a τ_p-relatively compact subset of E^X. Then*

$$\operatorname{ck}(H) \le \hat{d}(\overline{H}^{E^X}, C(X, E)) \le 2\operatorname{ck}(H).$$

In the particular case when $E = \mathbb{R}$, the space X can be taken normal and countably tight and we have

$$\hat{d}(\overline{H}^{\mathbb{R}^X}, C(X)) = \operatorname{ck}(H).$$

6. Baire one functions

It is known that when E is a Banach space, the uniform limits of Baire one functions are Baire one functions again. Hence, for a function $f \in E^X$ we have that $f \in B_1(X,E)$ if and only if $d(f, B_1(X,E)) = 0$. Consequently, for any subset $A \subset E^X$ we have $\hat{d}(A, B_1(X,E)) = 0$ if and only if $A \subset B_1(X,E)$. In this way and similarly to the case of continuous functions, when $E = \mathbb{R}$ and $H \subset \mathbb{R}^X$ is pointwise bounded, the number $\hat{d}(\overline{H}^{\mathbb{R}^X}, B_1(X))$ gives us a measure of non τ_p-compactness of H relative to $B_1(X)$ — observe that $\hat{d}(\overline{H}^{\mathbb{R}^X}, B_1(X)) = 0$ implies that H is τ_p-relatively compact in $B_1(X)$. Henceforth, we might now pursue the study we already did for continuous functions but now dealing with Baire one functions. In order to do so, the first difficulty to overcome is to answer to the following question:

Question 6.1. Given $f \in Z^X$, is there any way to estimate the distance $d(f, B_1(X,Z))$?

To effectively compute this distance, we use the concept of fragmented and σ-fragmented map as introduced in [24]. Recall that for a given $\varepsilon > 0$, a metric space–valued function $f \colon X \to (Z,d)$ is ε-fragmented if for each non-empty subset $F \subset X$ there exists an open subset $U \subset X$ such that $U \cap F \neq \emptyset$ and diam $(f(U \cap F)) \leq \varepsilon$. Given $\varepsilon > 0$, we say that f is ε-σ-fragmented by *closed sets* if there is a countable closed covering $(X_n)_n$ of X such that $f|_{X_n}$ is ε-fragmented for each $n \in \mathbb{N}$.

Definition 6.1. Let X be a topological space, (Z,d) a metric space and $f \in Z^X$ a function. We define:

$$\text{frag}\,(f) := \inf\{\varepsilon > 0 \colon f \text{ is } \varepsilon\text{-fragmented}\},$$

$$\sigma\text{-frag}_c(f) := \inf\{\varepsilon > 0 \colon f \text{ is } \varepsilon\text{-}\sigma\text{-fragmented by closed sets}\},$$

where, by definition, $\inf \emptyset = +\infty$.

The indexes frag and σ-frag$_c$ are related to each other as follows:

Theorem 6.1 ([5]). *Let X be a topological space and (Z,d) a metric space. If $f \in Z^X$ then the following inequality holds*

$$\sigma\text{-frag}_c(f) \leq \text{frag}\,(f).$$

If, moreover, X is hereditarily Baire, then

$$\sigma\text{-frag}_c(f) = \text{frag}\,(f).$$

With frag and σ-frag$_c$ one can estimate distances to $B_1(X,E)$.

Theorem 6.2 ([5]). *Let X be a metric space and E a Banach space. If $f \in E^X$ then*

$$\frac{1}{2}\,\sigma\text{-frag}_c(f) \le d(f, B_1(X,E)) \le \sigma\text{-frag}_c(f).$$

In the case $E = \mathbb{R}$, we have the equality

$$d(f, B_1(X)) = \frac{1}{2}\,\sigma\text{-frag}_c(f).$$

Next result is a consequence of the two previous ones.

Corollary 6.1 ([5]). *If X is a hereditarily Baire metric space and $f \in \mathbb{R}^X$, then*

$$d(f, B_1(X)) = \frac{1}{2}\text{frag }(f).$$

Note that the corollary above extends [20, Proposition 6.4], where this result is only proved when X is Polish.

Bearing in mind the definitions involved one proves:

Lemma 6.1 ([5]). *Let X be a separable metric space, (Z,d) a metric space and H a pointwise relatively compact subset of (Z^X, τ_p). Then (closures are taken relative to τ_p),*

$$\sup_{f\in\overline{H}} \text{frag }(f) = \sup_{\phi\in H^{\mathbb{N}}} \inf\{\text{frag }(f) \colon f \in \text{clust }(\phi)\}. \tag{10}$$

As we have done already in the case of continuous functions, we can study how far a set $H \subset E^X$ from being τ_p-relatively countably compact with respect to $B_1(X,E)$ using

$$\text{ck}_{B_1}(H) := \sup_{\varphi\in H^{\mathbb{N}}} d(\text{clust}_{Z^X}(\varphi), B_1(X,E)).$$

If we combine all the above, we can prove the following quantitative result about the difference between τ_p-relative compactness and τ_p-relative countable compactness with respect to $B_1(X,E)$. The particular case of ck $(H) = 0$ and $E = \mathbb{R}$ is the classic result due to Rosenthal ([29]).

Theorem 6.3 ([5]). *Let X be a Polish space, E a Banach space and H a τ_p-relatively compact subset of E^X. Then*

$$\text{ck }(H) \le \hat{d}(\overline{H}^{E^X}, B_1(X,E)) \le 2\text{ck }(H).$$

In the particular case when $E = \mathbb{R}$ we have

$$\hat{d}(\overline{H}^{\mathbb{R}^X}, B_1(X)) = \text{ck }(H).$$

7. Further studies

The very idea that *"qualitative"* properties can be derived from some *"inequalities"* is likely true for a great number of results. In our papers [2,5] there are more *"quantitative"* versions of classical results. We name some of them in the lines below.

In [5], we also obtain, with I. Namioka, a quantitative version of a Srivatsa's result that states that whenever X is metric any weakly continuous function $f \in E^X$ belongs to $B_1(X, E)$: our result here says that for an arbitrary $f \in E^X$ we have

$$d(f, B_1(X, E)) \leq 2 \sup_{x^* \in B_{E^*}} \text{osc}\,(x^* \circ f).$$

As a consequence, it is proved that for functions in two variables $f\colon X \times K \to \mathbb{R}$, X complete metric and K compact, there exists a G_δ-dense set $D \subset X$ such that the oscillation of f at each $(x, k) \in D \times K$ is bounded by the oscillations of the *partial* functions f_x and f^k. We indeed prove using games, that if X is a σ-β-unfavorable space and K is a compact space, then there exists a dense G_δ-subset D of X such that, for each $(y, k) \in D \times K$,

$$\text{osc}\,(f, (y, k)) \leq 6 \sup_{x \in X} \text{osc}\,(f_x) + 8 \sup_{k \in K} \text{osc}\,(f^k).$$

When the right hand side of the above inequality is zero, we are dealing with separately continuous functions $f\colon X \times K \to \mathbb{R}$ and we obtain as a particular case some well-known results obtained by I. Namioka in the mid 1970's.

The first named author has studied in [2] the distances from the set of selectors Sel (F) of a set–valued map $F\colon X \to \mathcal{P}(E)$ to the space $B_1(X, E)$. To do so, the notion of d-τ-semioscillation of a set–valued map with values in a topological space (Y, τ) also endowed with a metric d is introduced. Being more precise, it is proved that

$$d(\text{Sel}\,(F), B_1(X, E)) \leq 2\text{osc}^*_w(F)$$

where $\text{osc}^*_w(F)$ is the $\|\cdot\|$-w-semioscillation of F. In particular, when F takes closed values and $\text{osc}^*_w(F) = 0$ it is obtained that F has a Baire one selector: it should be pointed out that if F is weakly upper semicontinuous then $\text{osc}^*_w(F) = 0$ and therefore, these results strengthen a Srivatsa selection Theorem when F takes closed set.

More results along this line for other kind of spaces are foreseeable when studying distances to spaces of measurable functions, to spaces of integrable functions, etc. We are making an effort in this direction right now: if the results obtained are worth it, they will be published elsewhere.

Acknowledgements: The two authors are supported by the Spanish grants MTM2005-08379 (MEC and FEDER) and 00690/PI/04 (Fund. Séneca). The first named author is also supported by the FPU grant AP2003-4443 (MEC)

References

1. C. Angosto, *Distance to spaces of functions*, Ph. D. Thesis, Universidad de Murcia (2007).
2. C. Angosto, *Distances from selectors to spaces of Baire one functions*, Top. Appl. **155**(2) (2007), 69–81.
3. C. Angosto and B. Cascales, *The quantitative difference between countable compactness and compactness*, to appear in J. Math. Anal. Appl. (2008).
4. C. Angosto and B. Cascales, *Measures of weak noncompactness in Banach spaces*, to appear in Top. Appl. (2008).
5. C. Angosto, B. Cascales and I. Namioka, *Distances to spaces of Baire one functions*, preprint (2007).
6. A. V. Arkhangel'skiĭ, *Topological function spaces*, Mathematics and its Applications (Soviet Series), vol. 78, Kluwer Academic Publishers Group, Dordrecht (1992).
7. K. Astala and H. O. Tylli, *Seminorms related to weak compactness and to Tauberian operators*, Math. Proc. Cambridge Philos. Soc. **107**(2) (1990), 367–375.
8. J. Banaś and A. Martinón, *Measures of weak noncompactness in Banach sequence spaces*, Portugal. Math. **52**(2) (1995), 131–138.
9. Y. Benyamini and J. Lindenstrauss, *Geometric nonlinear functional analysis: Vol. 1*, American Mathematical Society Colloquium Publications, vol. 48, American Mathematical Society, Providence, RI (2000).
10. J. Bourgain, D. H. Fremlin and M. Talagrand, *Pointwise compact sets of Baire measurable functions*, Amer. J. Math. **100** (1978) 845–886.
11. B. Cascales, W. Marciszesky and M. Raja, *Distance to spaces of continuous functions*, Top. Appl., **153**(13) (2006), 2303–2319.
12. G. Choquet, *Lectures on analysis: Vol. II*, Benjamin, New York (1969).
13. F. S. De Blasi, *On a property of the unit sphere in a Banach space*, Colloq. Math. **65** (1992), 333–343.
14. J. Dieudonné and L. Schwartz, *La dualit dans les espaces (F) et (LF)*, Ann. Inst. Fourier Grenoble **1** (1949), 61–101.
15. W. F. Eberlein, *Weak compactness in Banach spaces*, Proc. Nat. Acad. Sci. U. S. A. **33**(1947), 51–53.
16. M. Fabian, P. Hájek, V. Montesinos and V. Zizler, *A quantitative version of Krein's theorem*, Rev. Mat. Iberoamericana **21**(1) (2005), 237–248.
17. K. Floret, *Weakly compact sets*, Lecture Notes in Mathematics, vol. 801, Springer, Berlin (1980).
18. A. S. Granero, *An extension of the Krein–Smulian Theorem*, Rev. Mat. Iberoamericana **22**(1) (2005) 93–110.

19. A. S. Granero, P. Hájek and V. Montesinos, *Convexity and w^*-compactness in Banach spaces*, Math. Ann. **328** (2004), 625–631.
20. A. S. Granero and M. Sánchez, *Convexity, compactness and distances*, Methods in Banach spaces, London Math. Soc. Lecture Note Ser. vol. 337, Cambridge Univ. Press, Cambridge (2006), 215–237.
21. A. S. Granero and M. Sánchez, *The class of universally Krein–Šmulian Banach spaces*, Bull. Lond. Math. Soc. **39** (2007), 529–540.
22. A. Grothendieck, *Critères de compacit dans les espaces fonctionnels gnraux*, Amer. J. Math. **74** (1952), 168–186.
23. G. J. Jameson, *Topology and normed spaces*, Chapman and Hall, London (1974).
24. J. E. Jayne, J. Orihuela, A. J. Pallarés and G. Vera, *σ-fragmentability of multivalued maps and selection theorems*, J. Funct. Anal. **117**(2) (1993), 243–273.
25. J. L. Kelley and I. Namioka, *Linear topological spaces*, Graduate Texts in Mathematics, vol. 36, Springer-Verlag, New York (1976).
26. A. Kryczka and S. Prus, *Measure of weak noncompactness under complex interpolation*, Studia Math. **147**(1) (2001), 89–102.
27. J. Orihuela, *Pointwise compactness in spaces of continuous functions*, J. London Math. Soc. (2) **36**(1) (1987), 143–152.
28. R. Pol, *On a question of H. H. Corson and some related problems*, Fund. Math. **109**(2) (1980), 143–154.
29. H. P. Rosenthal, *A characterization of Banach spaces containing ℓ^1*, Proc. Nat. Acad. Sci. U.S.A. **71** (1974), 2411–2413.
30. V. Šmulian, *Über lineare topologische Räume*, Rec. Math. (Mat. Sbornik) **7** (1940), 425–448.

SPACES OF SMOOTH FUNCTIONS

ELEONOR HARBOURE

Department of Mathematics
Wayne State University
48202 Detroit, MI, USA
E-mail: bert@math.wayne.edu

We will pose several situations in analysis where some classes of smooth functions play a fundamental role. In connection with the study of Laplace equation, we shall analyze the behavior of the fractional integral operator on L^p spaces, where BMO and Lipschitz spaces arise in a natural way. As a generalization, we will present and study a family of spaces introduced by Spanne. In particular we will be interested in identifying those members of the family containing only continuous functions. Finally we shall present a brief description of Besov spaces and their connection with a problem of non-linear approximation of a function by its wavelet expansion.

1. Fractional integration

The fractional integral operator arises in a natural way when solving problems involving differential operators. From elementary one variable calculus we know that integration and differentiation are inverse operations. This is basically the content of the Fundamental Theorem of Calculus. The picture is not that simple in higher dimensions where the most interesting situations occur. In order to solve partial differential equations, even in a theoretical framework, we must deal with operators involving inverses of "derivatives". Fractional integrals are in many cases the key operators to handle such inverses. The basic identity that leads to a generalization of the fundamental theorem of calculus in one variable, i.e., $\int_a^t f'(s)\,ds = f(t) - f(a)$, is the following:

$$f(x) = c_n \int_{\mathbb{R}^n} \frac{\langle \nabla f(y), x - y \rangle}{|x - y|^n}\, dy,$$

where f denotes a function defined on $\mathbb{R}^n$, with compact support and continuous partial derivatives. In fact, let $B(x, R)$ be a ball centered at x and with radius large enough to contain the support of f. For each unit direction

y' we may apply the one dimensional result to get

$$f(x) = \int_0^R D_{y'} f(x - ty')\, dt = \int_0^\infty \langle \nabla f(x - ty'), y' \rangle\, dt.$$

Integrating both sides over all the directions y' we obtain

$$f(x) = c_n \int_{S^{n-1}} \int_0^\infty \frac{\langle \nabla f(x - ty'), ty' \rangle}{t^n} t^{n-1}\, dt\, dy'$$

$$= c_n \int_{\mathbb{R}^n} \frac{\langle \nabla f(x - y), y \rangle}{|y|^n}\, dy = c_n \int_{\mathbb{R}^n} \frac{\langle \nabla f(y), x - y \rangle}{|x - y|^n}\, dy.$$

From here it follows immediately that

$$|f(x)| \leq c_n \int_{\mathbb{R}^n} \frac{|\nabla f(y)|}{|x - y|^{n-1}}\, dy.$$

Now we introduce the definition of the Fractional Integral operator of order α, $0 < \alpha < n$, by the expression

$$I_\alpha g(x) = \int_{\mathbb{R}^n} \frac{g(y)}{|x - y|^{n-\alpha}}\, dy.$$

It follows that, taking $\alpha = 1$ (as long as n is greater than one),

$$|f(x)| \leq c_n I_1(|\nabla f|).$$

As a consequence we may say that an improvement on the integrability of the function $I_\alpha(g)$ with respect to that of g, i.e., some boundedness results of I_α on Lebesgue spaces, would lead to obtain a better degree of integrability for a function f from assumptions on the size of its gradient. As an example, if we start with a function in L^2 whose gradient belongs also to L^2 and we are able to prove that the Fractional Integral operator for $\alpha = 1$ maps L^2 into L^q for some $q > 2$, we might conclude that f has in fact a better local integrability than that originally assumed. This type of result is known as one of the "immersion Sobolev's theorems" and it turns to be a fundamental tool in proving regularity properties for weak solutions to some uniformly elliptic partial differential equations, like the Laplace equation. In a similar way, results on the behavior of I_α over smooth function spaces are fundamental for obtaining regularity properties for classical solutions of such kind of equations. During the last fifty years, Fractional Integral operators have been intensively studied, not only in the present context, but in more general situations to englobe larger classes of equations.

Another way of looking at the relationship between Fractional Integral operators and derivatives is by studying their Fourier transforms. Since they are convolution operators, it is enough to know the Fourier transform of the kernel $k(x) = |z|^{\alpha-n}$. A homogeneity argument allows us to see that $\hat{k}(\xi)$ is, up to a constant, $|\xi|^{-\alpha}$. On the other hand, if we compute the Fourier transform of $(-\Delta)^{\alpha/2}$ using distributional calculus, we easily find that it is a constant times $|\xi|^{\alpha}$. Therefore, the composition of I_α with $(-\Delta)^{\alpha/2}$, whenever possible, gives the identity.

We shall start our study by stating some classical results concerning the behavior of these operators on the Lebesgue space $L^p(\mathbb{R}^n)$, that is, the set of measurable functions defined on $\mathbb{R}^n$ such that $|f|^p$ is integrable.

Theorem 1.1 (Hardy–Littlewood–Sobolev). *Let* $0 < \alpha < n$ *and* $1 < p < n/\alpha$. *Then* I_α *is a bounded operator from* $L^p(\mathbb{R}^n)$ *into* $L^q(\mathbb{R}^n)$ *with* $1/q = 1/p - \alpha/n$, *that is, there exists a constant* C *such that*

$$\|I_\alpha f\|_q \le C\|f\|_p.$$

Remark 1.1. It is a pleasant exercise to check that, because of the homogeneity of the kernel, if the operator I_α maps L^p into L^q, the relationship $1/q = 1/p - \alpha/n$ must hold. In fact, choosing g with $\|g\|_{L^p} = 1$, the above norm inequality applied to $f(x) = g(\lambda x)$ gives that

$$\lambda^{-\alpha-n/q} \le C\lambda^{-n/p}$$

should be true for any $\lambda > 0$. That is possible only if $\alpha + n/q = n/p$, which is the same as $1/q = 1/p - \alpha/n$.

In order to prove the theorem we first introduce a new space, a little bit larger than L^q, named weak-L^q or $L^{q,*}$ for short. Given a measurable function f, let us denote by μ_f its distribution function, that is, for $\lambda > 0$

$$\mu_f(\lambda) = |\{x\colon |f(x)| > \lambda\}|.$$

We will say $f \in L^{q,*}(\mathbb{R}^n)$ if there is a constant c such that

$$\mu_f(\lambda) \le \frac{c}{\lambda^q} \qquad \text{for all } \lambda > 0.$$

The infimum of such constants raised to $1/p$-th power turns to be a norm in this space as long as $1 \le q < \infty$ and, moreover, it is complete. The well–known Tchebycheff's inequality

$$\mu_f(\lambda) \le \frac{1}{\lambda^q}\int_{\mathbb{R}^n} |f|^q,$$

implies that $L^q \subset L^{q,*}$ continuously. On the other hand, it is straightforward to check that $g(x) = 1/|x|^{n/q}$ belongs to $L^{q,*}$ but, however, g does not belong to L^q.

Now, if a given operator T is bounded from L^p into $L^{q,*}$, we shall say that it is of weak type (p,q), while we shall say that T is of strong type (p,q), whenever it is bounded from L^p into L^q. From the above remark we deduce that any strong type operator is of weak type. However, the converse might be not true, as we shall illustrate later. We shall make use of a famous theorem due to Marcinkiewicz that will allow us to derive strong boundedness results from weak type inequalities. We give the precise statement (for a proof, see [6]).

Theorem 1.2 (Marcinkiewicz's interpolation theorem).
Let p_0, p_1, q_0, q_1 be real numbers such that $1 \leq p_i \leq q_i \leq \infty$, $p_0 < p_1$ and $q_0 \neq q_1$. Let T be a sublinear operator which is simultaneously of weak type (p_0, q_0) and (p_1, q_1). Then for each θ, $0 < \theta < 1$, with $1/p = (1-\theta)1/p_0 + \theta 1/p_1$ and $1/q = (1-\theta)1/q_0 + \theta 1/q_1$, we have that T is of strong type (p,q), that is,

$$\|Tf\|_q \leq A\|f\|_p.$$

(When $q_i = \infty$ weak type means $\|Tf\|_{q_i} \leq A_i\|f\|_{p_i}$.)

We shall also use the the very well–known Young's inequality for convolutions, namely

$$\|f * g\|_r \leq \|f\|_s\|g\|_t,$$

where $1 \leq s, t \leq \infty$ and $1 + 1/r = 1/s + 1/t$.

Proof of Theorem 1.1. We will prove that for $1 \leq p < n/\alpha$, $1/q = 1/p - \alpha/n$, the operator I_α satisfies

$$|\{x \colon |I_\alpha f|(x) > \lambda\}| \leq \frac{c}{\lambda^q}\left(\int |f|^p\right)^{q/p}.$$

In other words, I_α is of weak type $(p,q), 1 \leq p < n/\alpha$. From here, by means of Marcinkiewicz's interpolation theorem, we will obtain the strong type (p,q) in the range $1 < p < n/\alpha$.

For each $\eta > 0$, we split the kernel $K(x) = |x|^{\alpha - n}$ in

$$K = K_0 + K_\infty,$$

where $K_0 = K\chi_{B(0,\eta)}$ and $K_\infty = K\chi_{B^c(0,\eta)}$.

If f belongs to L^p, $K_0 * f$ as well as $K_\infty * f$ are finite a.e.. This is so since K_0 is an L^1 function while K_∞ is in $L^{p'}$, and then an application of Young's inequality gives that $K_0 * f$ belongs to L^p, and that $K_\infty * f$ is in

L^∞ and, consequently, finite almost everywhere. Moreover, straightforward calculations show that

$$\|K_0\|_1 \le c_0 \eta^\alpha, \quad \|K_\infty\|_{p'} = c_1 \eta^{-n/q}.$$

Now, let us observe that

$$\begin{aligned} &|\{x : |K * f|(x) > 2\lambda\}| \\ &\le |\{x\colon |K_0 * f|(x) > \lambda\}| + |\{x\colon |K_\infty * f|(x) > \lambda\}| \\ &= \mathrm{I} + \mathrm{II}. \end{aligned}$$

To estimate I, we use Tchebycheff's and Young's inequalities to get

$$\mathrm{I} \le \frac{1}{\lambda^p} \|K_0 * f\|_p^p \le \frac{1}{\lambda^p} \|K_0\|_1^p \|f\|_p^p \le c_0^p \left(\frac{\eta^\alpha \|f\|_p}{\lambda}\right)^p.$$

On the other hand, since for almost every x,

$$|K_\infty * f|(x) \le \|K_\infty\|_{p'} \|f\|_p \le c_1 \eta^{-n/q} \|f\|_p,$$

choosing η such that $c_1 \eta^{-n/q} \|f\|_p = \lambda$ we obtain that $\mathrm{II} = 0$.

Consequently, for this value of η we have

$$|\{x\colon |K * f|(x) > 2\lambda\}| \le \frac{c_0^p}{c_1^{pq\alpha/n}} \left(\frac{\|f\|^{q\alpha/n}}{\lambda^{q\alpha/n}} \frac{\|f\|_p}{\lambda}\right)^p = c \left(\frac{\|f\|_p}{\lambda}\right)^q,$$

since $1 + q\alpha/n = q/p$. □

Therefore we have shown that I_α is of weak type (p, q) for p in the interval $[1, n/\alpha)$. Since any p in the open interval $(1, n/\alpha)$ may be seen as an intermediate point between two values p_0 and p_1 belonging to the same interval, we may conclude via interpolation that I_α is of strong type for $p \in (1, n/\alpha)$ and q such that $1/q = 1/p - \alpha/n$.

Remark 1.2. In the above proof, we have seen that I_α is also of weak type (p, q) when $p = 1$ and $q = n/(n - \alpha)$.

Moreover it can be shown that it is not of strong type in the extreme point. In fact, if we take a sequence of functions f_k, $\|f_k\|_1 = 1$ tending to the Dirac delta we will have

$$I_\alpha f_k(x) = K * f_k \to c_n / |x|^{n-\alpha},$$

for almost every $x \in \mathbb{R}^n$. Therefore, if the strong type inequality were true we would have

$$\|K * f_k\|_{n/(n-\alpha)} \le A,$$

and by Fatou's Theorem, we would arrive to

$$\int_{\mathbb{R}^n} |x|^{-n}\, dx < \infty,$$

which is obviously false.

We state our observation as another boundedness result for I_α.

Theorem 1.3. *Let $0 < \alpha < n$. The operator I_α is of weak type $(1, n/(n-\alpha))$ but not of strong type.*

It is also not difficult to check that in the other end point $p = n/\alpha$, I_α is not of strong type $(n/\alpha, \infty)$ as may be expected. In this case it is enough to take $f(x) = |x|^{-\alpha}(\log 1/|x|)^{-r\alpha/n}\chi_{B(0,1/2)}(x)$, with $1 < r \le n/\alpha$, which belongs to $L^{n/\alpha}$ and observe that

$$I_\alpha f(x) = \int_{|y|\le 1/2} \frac{|y|^{-\alpha}}{|x-y|^{n-\alpha}} (\log 1/|y|)^{-r\alpha/n}\, dy,$$

is a continuous function for $x \neq 0$, and also

$$\lim_{x\to 0} I_\alpha f(x) = \int_{|y|\le 1/2} (\log 1/|y|)^{-r\alpha/n}|y|^{-n}\, dy = \infty,$$

since $1 - r\alpha/n \ge 0$, giving that $I_\alpha f$ is not essentially bounded. Then, a natural question arises. What can be said about $I_\alpha(f)$ for a function $f \in L^{n/\alpha}$? Certainly we should enlarge the space L^∞ so as to allow functions behaving locally as the logarithm at the origin. The appropriate space is known as *BMO* (bounded mean oscillation) or the John–Nirenberg space (see [2]) and it is defined as:

$$BMO = \left\{ f \in L^1_{\text{loc}} \colon \|f\|_* = \sup_B \frac{1}{|B|} \int_B |f(x) - m_B f|\, dx < \infty \right\},$$

where the supremum is taken over the family of balls in $\mathbb{R}^n$ and $m_B f$ denotes the average of f over the ball B, that is, $m_B f = |B|^{-1} \int_B f$.

If we want $\|\cdot\|_*$ to be a norm, we must identify those functions whose difference is a constant.

With this notation we will be able to prove the following result:

Theorem 1.4. *Let $f \in L^{n/\alpha}$ and having compact support. Then, $I_\alpha f$ is finite almost everywhere and*

$$\|I_\alpha f\|_* \le C\|f\|_{n/\alpha}.$$

Proof. Since such f belongs for instance to L^p, $1 < p < n/\alpha$, then $I_\alpha f \in L^q$, $1/q = 1/p - \alpha/n$, and hence, locally integrable.

Let $B = B(x_0, r)$ be a ball. We decompose $f = f_1 + f_2$ with $f_1 = f\chi_{\tilde{B}}$ where $\tilde{B} = B(x_0, 2r)$. Now,

$$\frac{1}{|B|}\int_B |I_\alpha f - m_B I_\alpha f| \leq \frac{2}{|B|}\int_B |I_\alpha f_1| + \frac{1}{|B|}\int_B |I_\alpha f_2 - m_B I_\alpha f_2| = \mathrm{I} + \mathrm{II}.$$

But if we choose p and q such that $1 < p < n/\alpha$, $1/q = 1/p - \alpha/n$, we obtain

$$\frac{1}{|B|}\int_B |I_\alpha f_1| \leq \left(\frac{1}{|B|}\int_B |I_\alpha f_1|^q\right)^{1/q} \leq C\frac{1}{|B|^{1/q}}\left(\int |f_1|^p\right)^{1/p},$$

in view of Theorem 1.1. Applying Hlder's inequality with $r = n/\alpha p > 1$ and $r' = n/(n - \alpha p)$ we get

$$\mathrm{I} \leq c\frac{|\tilde{B}|^{(n-\alpha p)/np}}{|B|^{1/q}}\|f\|_{n/\alpha} = c\|f\|_{n/\alpha}.$$

On the other hand

$$\mathrm{II} \leq \frac{1}{|B|^2}\int_B\int_B\int_{\tilde{B}^c} |f_2(y)|\left||x-y|^{\alpha-n} - |z-y|^{\alpha-n}\right| dy\, dz\, dx.$$

Since $x, z \in B$ and $y \in \tilde{B}^c$, $|x - y| \geq r$, $|z - y| \geq r$. An application of the mean value theorem leads to

$$\left||x-y|^{\alpha-n} - |z-y|^{\alpha-n}\right| \leq c|x - z|\theta^{\alpha-n-1},$$

being θ an intermediate value between $|x-y|$ and $|z-y|$. Since in our situation both values are equivalent to $|x_0 - y|$, the last expression is bounded by $cr|x_0 - y|^{\alpha-n-1}$, and then

$$\mathrm{II} \leq c \cdot r \int_{|x_0-y|>2r} |f(y)||x_0 - y|^{\alpha-n-1}\, dy$$

$$\leq c \cdot r\|f\|_{n/\alpha}\left(\int_{|x_0-y|>r} |x_0 - y|^{(\alpha-n-1)n/(n-\alpha)}\, dy\right)^{(n-\alpha)/n}.$$

Changing to polar coordinates the last integral equals to a constant times

$$\int_r^\infty \rho^{-n-n/(n-\alpha)}\rho^{n-1}\, d\rho = c \cdot r^{-n/(n-\alpha)},$$

and, therefore, we also obtain

$$\mathrm{II} \leq c\|f\|_{n/\alpha}. \qquad \square$$

Remark 1.3. We have stated the theorem only for $I_\alpha f$ with f in $L^{n/\alpha}$ and having compact support. Let us notice that for such functions, if we define

$$\tilde{I}_\alpha f(x) = \int_{\mathbb{R}^n} \left(\frac{1}{|x-y|^{n-\alpha}} - \frac{\chi_{B^c(0,1)}}{|y|^{n-\alpha}} \right) f(y)\, dy$$

$$= I_\alpha f(x) - \int_{|y|\geq 1} \frac{f(y)}{|y|^{n-\alpha}}\, dy = I_\alpha f(x) - C,$$

we would obtain that $\tilde{I}_\alpha f$ and $I_\alpha f$ are the same as functions in BMO. On the other hand, it is easy to see that for $f \in L^{n/\alpha}$, $\tilde{I}_\alpha f$ is finite almost everywhere and, moreover, locally integrable. In fact, let $B_R = B(0,R)$ with $R > 1$ and $x \in B_R$. We write

$$\tilde{I}_\alpha f(x) = \int_{|y|\leq 2R} \frac{f(y)}{|x-y|^{n-\alpha}}\, dy + \int_{1\leq |y|\leq 2R} \frac{f(y)}{|y|^{n-\alpha}}\, dy$$

$$+ \int_{|y|\geq 2R} \left[\frac{1}{|x-y|^{n-\alpha}} - \frac{1}{|y|^{n-\alpha}} \right] f(y)\, dy.$$

The first term in the sum gives a function in L^1_{loc} since it is the fractional integral of a $L^{n/\alpha}$ function with compact support. The second integral is a finite quantity and independent of x since $1/|y|^{n-\alpha} \leq 1$ and, being f in $L^{n/\alpha}$, is locally integrable. Finally, for x in B_R, the quantity between brackets is a difference of two values of the function $t^{\alpha-n}$ away from the origin, and hence the mean value theorem may be applied to bound the integrand by $C|x|/|y|^{n-\alpha+1}$, since again $|x-y| \simeq |y|$. Clearly, this last function belongs to $L^{n/(n-\alpha)}$, and then Hlder's inequality gives that the third integral is bounded by $C|x|$ which is integrable on B_R.

From these observations we can say that $\tilde{I}_\alpha f$ provides an extension of $I_\alpha f$ for general functions belonging to $L^{n/\alpha}$ and not necessarily with compact support.

Similar considerations hold for $n/\alpha \leq p < n/(\alpha-1)^+$. It turns out that $\tilde{I}_\alpha f$ is also well defined for $f \in L^p$, giving a class of locally integrable functions that differ by a constant. In fact the same argument applies and all we need is that $1/|y|^{n-\alpha} \in L^{p'}_{\text{loc}}$ and $1/|y|^{n-\alpha+1}\chi_{B^c_R} \in L^{p'}$, and clearly both are true in the stated range.

A new question therefore arises: what can be said about the image of L^p under $\tilde{I}_\alpha$ when $n/\alpha < p < n/(\alpha-1)^+$?

From the above remark we know that $\tilde{I}_\alpha f$ is locally integrable and the proof of Theorem 1.3 can be followed step by step; the only difference is that when estimating the averages in terms of $\|f\|_p$ instead of $\|f\|_{n/\alpha}$, we will obtain $C\|f\|_p|B|^{\alpha/n-1/p}$ on the right hand side.

In this way we would get an estimate of the type

$$\frac{1}{|B|^{\alpha/n-1/p}}\frac{1}{|B|}\int_B |\tilde{I}_\alpha f - m_B\tilde{I}_\alpha f| \leq C\|f\|_p,$$

for p such that $n/\alpha < p < n/(\alpha-1)^+$. Let us observe that, in such a situation, the exponent $\alpha/n-1/p$ is always positive and less than $1/n$. Moreover, the above inequality for $p = n/\alpha$ gives the statement of Theorem 1.3.

Then, for a given $0 \leq \beta < 1$ we introduce the space

$$BMO_\beta = \left\{f \in L^1_{\text{loc}} \colon \sup_B \frac{1}{|B|^{\beta/n}}\frac{1}{|B|}\int_B |f - m_B f| < \infty\right\}.$$

When $\beta = 0$, we recover *BMO* and for $\beta > 0$, as we shall see in the next section, it coincides with a very well–known space of smooth functions.

2. Functions with controlled mean oscillation

As a generalization of the spaces we just introduced, when trying to describe the image of L^p $(p > n/\alpha)$ under the fractional integration, S. Spanne ([5]) defined the BMO_φ spaces as the class of functions whose mean oscillation is controlled by φ, a fixed non-decreasing and positive function defined on $(0,\infty)$. More precisely,

$$BMO_\varphi = \left\{f \in L^1_{\text{loc}} \colon \sup_B \frac{1}{\varphi(|B|^{1/n})}\frac{1}{|B|}\int_B |f - m_B f| < \infty\right\}$$

and, moreover, if we denote by $\|\cdot\|_{*,\varphi}$ that supremum taking over all the balls in $\mathbb{R}^n$,the space BMO_φ, turns to be a Banach space, after identifying those functions that differ by a constant.

Clearly for $\varphi(t) = t^\beta$, $0 \leq \beta < 1$, we have the spaces introduced in the previous section. In particular, for $\beta = 0$, we recover the John–Nirenberg space. These spaces were firstly studied by Campanato ([1]) and Meyers ([4]) in connection with the study of regularity of solutions of elliptic partial differential equations.

In this section, we plan to study some properties of these spaces. In particular, it is obvious that *BMO* $(\beta = 0)$ contains non continuous functions (obviously $L^\infty \subset BMO$), while in [1] and [4] it is shown that for $0 < \beta < 1$, all the functions are continuous and, moreover, their modulus of continuity is not worse than t^β.

Spanne ([5]) considered the problem of smoothness for functions in BMO_φ, posing the questions of finding conditions on φ to guarantee that

BMO_φ contains only smooth functions and when such situation does not occur.

To answer these questions we introduce the space of Lipschitz-φ functions, as those functions whose modulus of continuity is controlled by φ, i.e.,

$$\Lambda_\varphi = \left\{ f : \omega_f(t) = \sup_{|x-y|\le t} |f(x) - f(y)| \le c\varphi(t) \right\}.$$

It is immediate to check that $\Lambda_\varphi \subset BMO_\varphi$ and also that $\Lambda_\varphi = L^\infty/c$ when $\varphi(t) \simeq 1$ (here, L^∞/c means that we have identified functions differing a.e. by a constant.)

In the next theorem, we state the results by Spanne.

Theorem 2.1. *Let φ be a non-decreasing and positive function. Then we have*

(a) If the function φ also satisfies $\int_0^\delta \varphi(t)\,dt/t < \infty$ for some $\delta > 0$ then any function in BMO_φ is continuous and, moreover, $\omega_f(s) \le c\int_0^s \varphi(t)\,dt/t$.

(b) If $\varphi(t)/t$ is non increasing and $\int_0^\delta \varphi(t)\,dt/t$ diverges, then the space BMO_φ contains discontinuous and locally unbounded functions.

Corollary 2.1. *If φ is such that $\int_0^s \varphi(t)\,dt/t < \infty$ for some $\delta > 0$, denoting by $\tilde\varphi(s) = \int_0^s \varphi(t)\,dt/t$, it follows that $BMO_\varphi \subset \Lambda_{\tilde\varphi}$.*

We will not show the result (a) of Spanne in its full generality. Instead, to make the computations easier, we are going to assume that $\varphi(t)/t$ is non increasing also for the proof of (a).

We shall make use of the following simple lemma.

Lemma 2.1. *Let $f \in BMO_\varphi$ and $B \subset \overline{B}$ two balls in $\mathbb{R}^n$. Then*

$$|m_B f - m_{\overline{B}} f| \le \|f\|_{*,\varphi} \frac{|\overline{B}|}{|B|} \varphi(|\overline{B}|^{1/n}).$$

Proof.

$$|m_B f - m_{\overline{B}} f| = \frac{1}{|B|} \int_B |f - m_{\overline{B}} f|$$

$$\le \frac{|\overline{B}|}{|B|} \frac{1}{|\overline{B}|} \int_{\overline{B}} |f - m_{\overline{B}} f| \le \frac{|\overline{B}|}{|B|} \varphi(|\overline{B}|^{1/n}) \|f\|_{*,\varphi}. \qquad \square$$

Proof of theorem 2.1(a). Let us start by noticing that $\varphi(t)/t$ non increasing implies that for any fixed $a \ge 1$, there is a constant c such that

$\varphi(at) \leq c\varphi(t)$. On the other hand, if $a < 1$, such inequality holds with constant one, since φ is non-decreasing. It is also clear that $\varphi(t/2) \leq c\int_{t/2}^{t} \varphi(s)\,ds/s \leq c\int_0^t \varphi(s)\,ds/s$. Hence, $\varphi(t) \leq c\tilde{\varphi}(t)$.

Let $x, y \in \mathbb{R}^n$ and $B = B(x, |x-y|), B' = B(y, |x-y|)$ and $\tilde{B} = B(x, 2|x-y|)$.

$$|f(x) - f(y)|$$

$$\leq |f(x) - m_B f| + |f(y) - m_{B'} f| + |m_{B'} f - m_{\tilde{B}} f| + |m_{\tilde{B}} f - m_B f|$$

$$= \mathrm{I} + \mathrm{II} + \mathrm{III} + \mathrm{IV}.$$

Since both, B y B', are contained in $\tilde{B}$, the terms III and IV, according to Lemma 2.1, are bounded by

$$2^n \varphi(|\tilde{B}|^{1/n}) \|f\|_{*,\varphi} \leq c 2^n \varphi(|x-y|) \|f\|_{*,\varphi} \leq c\|f\|_{*,\varphi} \int_0^{|x-y|} \varphi(t)\,\frac{dt}{t}.$$

The terms I y II are quite similar, so we only bound the first. We set $B_i = B(x, 2^{-i}|x-y|)$ for $i \geq 1$ y $B_0 = B$. Then we have

$$|f(x) - m_B f| \leq |f(x) - m_{B_m} f| + \sum_{i=0}^{m-1} |m_{B_{i+1}} f - m_{B_i} f|.$$

Since f is locally integrable, Lebesgue's differentiation theorem applies. Let us assume that x is in fact a Lebesgue point. Then, taking limit for $m \to \infty$, the first term on the right hand side goes to zero, and applying Lemma 2.1 to each term in the series we get

$$|f(x) - m_B f| \leq \sum_{i=0}^{\infty} |m_{B_{i+1}} f - m_{B_i} f| \leq c\|f\|_{*,\varphi} \sum_{i=0}^{\infty} \varphi(2^{-i}|B|^{1/n})$$

$$\leq C'\|f\|_{*,\varphi} \sum_{i=0}^{\infty} \int_{2^{-i}}^{2^{-i+1}} \varphi(t|B|^{1/n})\,\frac{dt}{t} \leq C\|f\|_{*,\varphi} \int_0^1 \varphi(t|B|^{1/n})\,\frac{dt}{t}.$$

Since $|B|^{1/n} = \omega_n|x-y|$ with $\omega_n = |B(0,1)|^{1/n}$, performing the change of variables $s = t|x-y|$ and using that $\varphi(as) \leq c\varphi(s)$ it follows that

$$\mathrm{I} \leq c\|f\|_{*,\varphi} \int_0^{|x-y|} \varphi(s)\,\frac{ds}{s},$$

for some appropriate constant c. Therefore, (a) of the theorem is proved under the extra assumption $\varphi(t)/t$ non-increasing. $\square$

Before proceeding with the proof of (b) of the theorem, let us observe that a function f satisfying the property: for any ball B there is a constant C_B such that

$$\frac{1}{|B|}\int_B |f - C_B| \leq A\varphi(|B|^{1/n}),$$

with A independent of the ball B, certainly belongs to BMO_φ and, moreover, $\|f\|_{*,\varphi} \leq 2A$. In fact,

$$\frac{1}{|B|}\int_B |f - m_B f| \leq \frac{1}{|B|}\int_B |f - C_B| + |C_B - m_B f|$$

$$\leq A\varphi(|B|^{1/n}) + \frac{1}{|B|}\int_B |f - C_B| \leq 2A\varphi(|B|^{1/n}).$$

Consequently, in order to prove that a function does belong to BMO_φ, we may use any constant C_B instead of $m_B f$.

Proof of theorem 2.1(b). We set

$$h(x) = \int_{|x|}^{1} \frac{\varphi(t)}{t}\, dt.$$

Then h is continuous at $x \neq 0$ and under the assumptions on φ, it is discontinuous at $x = 0$ and unbounded nearby. To check that $h \in BMO_\varphi$, it is enough to consider balls $B(z,r)$ with $z \neq 0$.

We set $B = B(z,r)$, $z_B = z + r\frac{z}{|z|}$ and $C_B = h(z_B)$. Let us notice that $|z_B| = |z| + r$ and that for $x \in B$, $|x| \leq |x - z| + |z| \leq |z| + r$. Hence, for $x \in B$,

$$|h(x) - C_B| = |h(x) - h(z_B)| = \int_{|x|}^{|z|+r} \varphi(t)\,\frac{dt}{t}.$$

In order to estimate the oscillation, let us consider first the case $|z| < 2r$. In this situation we have

$$\int_B |h(x) - C_B| = \int_B \int_{|x|}^{|z|+r} \varphi(t)\,\frac{dt}{t}\,dx \leq \int_0^{|z|+r} \frac{\varphi(t)}{t}\left(\int_{|x|\leq t} dx\right) dt$$

$$\leq C\varphi(|z| + r)\int_0^{|z|+r} t^{n-1} \leq C\varphi(3r)\,r^n \leq C\varphi(|B|^{1/n})|B|,$$

where we have used φ non-decreasing, $|z| < 2r$ and $\varphi(ar) \leq C\varphi(r)$.

Now, if $|z| > 2r$, the distance from the origin to the ball is at least r. In fact, if $x \in B$, $|x| \geq |z| - |z - x| \geq |z| - r \geq r$. In this way

$$\int_B |h(x) - C_B| \leq \int_B \left(\int_r^{|z|+r} \varphi(t) \, \frac{dt}{t} \right) dx \leq |B| \frac{\varphi(r)}{r} 2r = 2|B|\varphi(|B|^{1/n}),$$

where we have used that $\varphi(t)/t$ is non-increasing. □

Remark 2.1.

(1) It is worth noting that the proof of $h \in BMO_\varphi$ does not make use of the divergence of the integral; we just used φ non-decreasing and $\varphi(t)/t$ non increasing.

(2) For $\varphi(t)/t$ non-increasing, (a) and (b) imply that if $\int_0^\delta \varphi(t)\, dt/t$ diverges then $\Lambda_\varphi \subsetneq BMO_\varphi$. In fact, when $\varphi(0^+) = 0$, all the functions in Λ_φ are continuous and when $\varphi(0^+) > 0$, they are bounded (locally). On the other hand, if the integral converges and $\varphi \simeq \tilde{\varphi}$, then $\Lambda_\varphi = BMO_\varphi$. Conversely, it can be seen that if both spaces agree, not only the integral must converge (a consequence of (b)) but $\varphi \simeq \tilde{\varphi}$ must hold. Indeed, by the previous remark, $h \in BMO_\varphi$ and hence $h \in \Lambda_\varphi$. Therefore,

$$|h(x) - h(0)| \leq C\varphi(|x|).$$

But, according to the definition of h,

$$|h(x) - h(0)| = \int_0^{|x|} \varphi(t) \, \frac{dt}{t} = \tilde{\varphi}(|x|).$$

Then $\tilde{\varphi}(r) \leq C\varphi(r)$ for any positive r. Since the converse inequality always holds, we arrive to $\varphi \simeq \tilde{\varphi}$.

(3) An example where the assumptions made in (b) hold is $\varphi(t) \equiv 1$. In such case the function h is

$$h(|x|) = \log(1/|x|),$$

which is the classical example of unbounded function (even locally) which does belong to BMO.

3. Smooth function spaces and wavelets

Besides BMO_φ, there are other families of spaces that generalize Lipschitz-α spaces. We will introduce another line of spaces and we shall present a problem arising in non-linear approximation where they become the appropriate spaces. We will follow closely the exposition given in the book by Wojtaszczyk ([7, Chapter 9]).

In the sequel, for simplicity, we will restrict our functions to one dimension, even though most of the results have an extension to higher dimensions.

As we have seen, a Lipschitz function is defined in terms of its pointwise modulus of continuity, i.e.,

$$\omega_f(t) = \sup_{|h|\leq t} \sup_x |f(x+h) - f(x)| = \sup_{|h|\leq t} \|f(x+h) - f(x)\|_\infty,$$

which measures, in some sense, the size of the difference between a function and its translation. Since there are many ways of measuring the size of a function, it is natural to introduce the p-modulus of continuity by

$$\omega_p(f,t) = \sup_{|h|\leq t} \|f(x+h) - f(x)\|_p.$$

Clearly, $\omega_\infty(f,t) = \omega_f(t)$.

Next we establish several simple properties of ω_p:

(i) $\omega_p(f,t)$ is a non-decreasing function of t.
(ii) If $1 \leq p < \infty$ and $f \in L^p$, then $\lim_{t\to 0} \omega_p(f,t) = 0$, and moreover $\omega_p(f,t) \leq 2\|f\|_p$ for $t > 0$.
(iii) $\omega_p(f,mt) \leq m\omega_p(f,t)$ if $m \in \mathbb{N}$.
(iv) $\lim_{t\to 0} t^{-1}\omega_p(f,t) = 0 \Rightarrow f = \text{constant}$.

Clearly, (i) and (iii) hold. For (ii), the claim on the limit is obvious for smooth functions with compact support, and the result follows by the density of such functions in L^p. Finally, from (iii) we get

$$\omega_p(f,t) = \omega_p(f,mt/m) \leq \frac{\omega_p(f,t/m)}{t/m}t.$$

Making m tend to infinity, the assumption in (iv) implies that $\omega_p(f,t) = 0$ for each $t > 0$ and then f equal a.e to a constant.

A function belongs to a Lipschitz-α space whenever $\sup_{t>0} t^{-\alpha}\omega_\infty(f,t) < \infty$. Again, we may change the sup–norm by a different norm, to introduce a new family of spaces: the non-homogeneous Besov spaces $\dot{B}^p_{\alpha,s}$, with $0 < \alpha \leq 1$ and $1 \leq p, s \leq \infty$, as the set of functions such that $\|f\|_{p,\alpha,s} < \infty$, where

$$\|f\|_{p,\alpha,s} = \begin{cases} \left(\int_0^\infty \left[t^{-\alpha}\omega_p(f,t)\right]^s \frac{dt}{t}\right)^{1/s} & \text{if } 1 \leq s < \infty, \\ \sup_{t>0} t^{-\alpha}\omega_p(f,t) & \text{if } s = \infty. \end{cases}$$

In fact, $\|\cdot\|_{p,\alpha,s}$ are seminorms and they vanish on constant functions. If we want to work with a norm we should identify functions differing by a constant, but the resulting spaces may not be complete. When these spaces are completed, they involve not only functions but also distributions, and their treatment becomes more difficult.

One way to bypass this difficulty is to introduce the so called homogeneous Besov spaces $B^p_{\alpha,s}$, as the set of functions in L^p such that $\|f\|_{p,\alpha,s} < \infty$. In this way, it turns to be a Banach space with respect to the norm $\|f\|_p + \|f\|_{p,\alpha,s}$.

The seminorm $\|\cdot\|_{p,\alpha,s}$ has a discrete version as it is easy to check.

Proposition 3.1. *For p, α, s as above, there exist positive constants c and C such that*

$$c\,\|f\|_{p,\alpha,s} \le \sum_{j\in\mathbb{Z}} 2^{\alpha j s}\omega_p(f, 2^{-j})^s \le C\|f\|_{p,\alpha,s}.$$

Proof. Splitting the integral into dyadic intervals and using (i) and (iii), we get

$$\int_0^\infty \left[t^{-\alpha}\omega_p(f,t)\right]^s \frac{dt}{t} = \sum_{j\in\mathbb{Z}} \int_{2^{-j}}^{2^{-j+1}} \omega_p(f,t)^s \frac{dt}{t^{\alpha s+1}} \le 2^s \sum_{j\in\mathbb{Z}} 2^{\alpha j s}\omega_p(f, 2^{-j})^s.$$

Also by (i), the integral is bounded below by

$$2^{-\alpha s-1}\sum_{j\in\mathbb{Z}} 2^{\alpha j s}\omega_p(f,2^{-j})^s$$

and the proposition is proved for $s < \infty$. The case $s = \infty$ follows similarly, just replacing integrals and sums by suprema. □

To illustrate a situation where Besov's spaces appear in a natural way, we shall introduce, in an informal way, some basics concepts and facts from the one dimensional wavelet theory.

A *wavelet* on $\mathbb{R}$ is a function $\psi \in L^2(\mathbb{R})$ such that the family of functions

$$\psi_{jk}(t) = 2^{j/2}\psi(2^j t - k) \quad \text{with } j, k \in \mathbb{Z}$$

gives an orthonormal system in $L^2(\mathbb{R})$.

A first natural question is whether or not such functions do exist. Let us observe that taking $\psi = \chi_{(0,1/2)} - \chi_{(1/2,1)}$, the family $\{\psi_{jk}\}$ is the well known Haar system that is in fact a basis for $L^2(\mathbb{R})$. If we want to have a wavelet ψ, smooth and with some decay at infinity, the examples are not that easy. Daubechies and Meyer, among others, constructed wavelets with both properties through a method called a multiresolution analysis.

One of the main advantages in analyzing functions by means of wavelets rather than through the Fourier method is that it makes possible to obtain characterizations of most of the useful function spaces in terms of wavelet coefficients.

Here we will not give details on what a multiresolution analysis is. For those knowing this method for constructing wavelets, we say that we will be working with a ψ coming from a scale–function ϕ satisfying

(i) $\phi \in C^1(\mathbb{R})$, and
(ii) $|\phi(x)| + |\phi'(x)| \le C(1+|x|)^{-A}$, with $A > 3$.

Given a function f, its coefficients with respect to the system ψ_{jk} are given by

$$\langle f, \psi_{jk}\rangle = \int f(t)\psi_{jk}(t)\, dt.$$

Although any function in L^2 can be described in terms of coefficients derived from any basis, if we have a system coming from a "good" wavelet, that result can also be extended to L^p, $1 < p < \infty$. In that case, we obtain

$$\|f\|_p \simeq \left\| \left(\sum |\langle f, \psi_{jk}\rangle|^2 \chi_{I_{jk}} |I_{jk}|^{-1} \right)^{1/2} \right\|_p ,$$

where I_{jk} denotes the interval $\left[k2^{-j}, (k+1)2^{-j}\right]$.

Also, the Besov seminorm of a function can be described in terms of wavelet coefficients. For the application we have in mind, we shall need the following result (for a proof, see [7, p. 228]).

Theorem 3.1. *Let ψ be a wavelet associated to a multiresolution analysis satisfying* (i) *and* (ii). *Assume further that $|\psi(x)| \le c(1+|x|)^{-A}$. Then, for $0 < \alpha < 1$ and $1 \le p, s \le \infty$, there exists a constant C such that*

$$\left(\sum_{j \in \mathbb{Z}} \left[2^{j\alpha} \left(\sum_k 2^{jp(1/2-1/p)} |\langle f, \psi_{jk}\rangle|^p \right)^{1/p} \right]^s \right)^{1/s} \le C\|f\|_{p,\alpha,s}.$$

We are interested in the case $s = p = (\alpha + 1/2)^{-1}$ with $\alpha \le 1/2$. In that situation the above theorem establishes

$$\sum_{j,k} |\langle f, \psi_{jk}\rangle|^p \le C|f\|^p_{p,\alpha,p}.$$

Assuming this result, it is our intention to investigate the following problem in data compression.

Suppose we have a function $f \in L^2$. We know in this case that f may be approximated in the L^2-norm by a finite sum of its expansion, $\sum_{jk}\langle f, \psi_{jk}\rangle\psi_{jk}$. Now, assume that we can keep records of only a fix number of coefficients N, not necessarily the first ones. How good is this approximation measured in the L^2-norm? In other words: can we express the order of the approximation in terms of N for all functions in L^2?

The following example shows that the answer is negative if we deal with a general function in L^2. In fact, suppose we are allowed to use N coefficients to approximate f, i.e., we search for the best approximation of f, in the sense of L^2, by $\sum_{(j,k)\in A}\langle f, \psi_{jk}\rangle\psi_{jk}$, where $A \subset \mathbb{Z} \times \mathbb{Z}$ and card $(A) \leq N$.

Let $f_N = \sum_{k=1}^{2N} \frac{1}{\sqrt{2N}}\psi_{0,k}$. Then $\|f_N\|_2 = 1$ and for any A with card $(A) \leq N$ we have

$$\left\| f_N - \sum_{(j,k)\in A} \langle f, \psi_{jk}\rangle\psi_{jk} \right\|_2 \geq \frac{1}{\sqrt{2}}.$$

This is so because the best choice for A is to keep non-vanishing coefficients and having only $2N$ of them and with the same size, we may choose for example $A = \{(0,k) \colon k = 1, \dots, N\}$ and the norm of the difference gives in this case $\left(\sum_{N+1}^{2N} 1/(2N)\right)^{1/2} = 1/\sqrt{2}$.

Since the L^2-norm of a function is the ℓ^2-norm of its coefficients taken with respect to an orthonormal basis, we may think the above problem in the following way.

Given a sequence $a = \{a_k\}_{k\in\mathbb{Z}}$, with $\|a\|_{\ell^2} = 1$ and a natural number N, what additional conditions on the sequence would guarantee that choosing the N largest coefficients (in absolute value) we will get a "good" approximation of the original one? ("good" means here that the error goes to zero with N, or better yet, that goes to zero like a negative power of N).

Let us define the set $B \subset \mathbb{Z}$ such that card $(B) = N$ and $|a_k| \geq |a_\ell|$ whenever $k \in B$ and $\ell \notin B$. Let b the sequence defined by

$$b_k = \begin{cases} a_k & \text{if } k \in B, \\ 0 & \text{if } k \notin B \end{cases}$$

Assume further that $a \in \ell^p$ for some p, $1 \leq p < 2$ with $\|a\|_{\ell^p} \leq C$. Then we have

$$\sup_{k\notin B}|a_k| \leq \min_{k\in B}|a_k| \leq \left(\frac{1}{N}\sum_{k\in B}|a_k|^p\right)^{1/p} \leq CN^{-1/p}.$$

Since $p<2$, it follows that

$$\begin{aligned}\|b-a\|_{\ell^2} &= \left(\sum_{k\notin B}|a_k|^2\right)^{1/2} = \left(\sum_{k\notin B}|a_k|^{2-p}|a_k|^p\right)^{1/2} \\ &\leq \left(CN^{-1/p}\right)^{1-p/2}C^{p/2} = CN^{1/2-1/p},\end{aligned}$$

and we obtain a "good" approximation since $1/2-1/p<0$.

We may rephrase what we have done in the following way. Assume as above $a\in\ell^2\cap\ell^p$ with $p<2$, $\|a\|_{\ell^2}=1$ and $\|a\|_{\ell^p}\leq C$. Instead of fixing N, we fix a lower threshold for the size of the coefficients, say δ with $\delta>0$, and let us approximate by the sequence neglecting those coefficients less than δ. Let now $B=\{k\colon |a_k|\geq\delta\}$ and define the sequence b as above. The previous estimates give that

$$\text{card}\,(B) \leq \frac{C^p}{\min_{k\in B}|a_k|^p} \leq C^p\delta^{-p},$$

and hence

$$\|b-a\|_{\ell^2} \leq C^{p/2}\,\delta^{1-p/2}.$$

In this way, the approximation improves as $\delta\to 0$, and the velocity of convergence increases when p gets closer to 1.

Coming back to wavelet expansions, the above discussion shows that, although such non linear approximation methods may not be good for all the functions in L^2, they will work for some special subspaces, namely those functions satisfying

$$\left(\sum_{j,k}|\langle f,\psi_{jk}\rangle|^p\right)^{1/p} \leq C$$

for some $p<2$.

The description of Besov spaces in terms of wavelet coefficients allows us to conclude that they are the appropriate spaces to make these methods converge. The precise result is the following.

Theorem 3.2. *Let* $0 < \alpha \leq 1/2$, $p = (\alpha + \frac{1}{2})^{-1}$ *and* $f \in L^2 \cap \dot{B}^p_{\alpha,p}$. *Then, there exists a constant* K *such that for any* $N \in \mathbb{N}$ *it is possible to find a set* A, $A \subset \mathbb{Z} \times \mathbb{Z}$ *with* card $(A) = N$ *and*

$$\left\| f - \sum_{(j,k)\in A} \langle f, \psi_{jk}\rangle \psi_{jk} \right\|_2 \leq K\|f\|_{p,\alpha,p} N^{-\alpha}.$$

Or, alternatively, there exists a constant M *such that for any* $\delta > 0$, *if* $B_\delta = \{(j,k)\colon |\langle f, \psi_{jk}\rangle| \geq \delta\}$, *we have*

$$\left\| f - \sum_{(j,k)\in B_\delta} \langle f, \psi_{jk}\rangle \psi_{jk} \right\|_2 \leq M\|f\|_{p,\alpha,p} \delta^{2\alpha}.$$

References

1. S. Campanato, *Propiet di hlderianit di alcune classi di funcioni*, Ann. Sc. Normale Sup. Pisa **17** (1963), 175–188.
2. F. John and L. Nirenberg, *On functions of bounded mean oscillation*, Comm. Pure Appl. Math. **14** (1961), 415–426.
3. E. Hernndez and G. Weiss, *A first course on wavelets*, CRC–Press (1996).
4. G. N. Meyers, *Mean oscillation over cubes and Hlder continuity*, Proc. Amer. Math. Soc **15** (1964), 717–721.
5. S. Spanne, *Some function spaces defined using the mean oscillation over cubes*, Ann. Sc. Normale Sup. Pisa **19** (1965), 593–608.
6. E. M. Stein, *Singular integrals and differentiability properties of functions*, Princeton University Press (1970).
7. P. Wojtaszczyk, *A mathematical introduction to wavelets*, London Mathematical Society Student Texts **37**, Cambridge University Press (1997).

DOMINATION BY POSITIVE OPERATORS AND STRICT SINGULARITY

FRANCISCO L. HERNÁNDEZ

Department of Mathematical Analysis
Universidad Complutense de Madrid
20840 Madrid, Spain
E-mail: pacoh@mat.ucm.es

The aim of this talk is to study the domination problem for the class of strictly singular operators and other related operator classes. We also discuss the strict singularity of inclusions between rearrangement invariant function spaces.

A central question in the theory of Positive Operators between Banach lattices is the so-called *domination* problem:

Let R and T be positive operators between the Banach lattices E and F such that $0 \leq R \leq T\colon E \to F$. Assume that T satisfies certain property $(*)$.

(i) Does the operator R inherit the property $(*)$?
(ii) What effect does the property $(*)$ have on the dominated operator R?

It can happen that for some properties the answer is the best possible: (i) has a positive answer. For example, the case for $(*)$ to be an integral operator, i.e., representable as

$$Tf(x) = \int_\Omega K(x,y)\, f(y)\, d\mu(y).$$

But in general, the answer is negative and thus the problem is to determine the weakest conditions on the involved Banach lattices for a positive answer.

The domination problem for the important class of compact operators was solved by P. Dodds and D. Fremlin in [7]:

Let E and F be Banach lattices with E^* and F order continuous *and consider operators* $0 \leq R \leq T \colon E \to F$*. If* T *is compact then* R *is also compact.*

In the special setting of $L^p(\mu)$-spaces, it was also solved independently by L. Pitt in [20]: If $1 < p \leq \infty$, $1 \leq q < \infty$ and we consider operators $0 \leq R \leq T \colon L^p(\mu) \longrightarrow L^q(\nu)$ with T compact then R is also compact. This compactness domination result have been applied in several areas like operator semigroup, ergodic theory, transport theory and bio–mathematica (see the survey [2] for references).

The domination for weakly compact operators was considered by A. Wickstead in [23] showing that if E^* or F are order continuous and T is weakly compact then R is also weakly compact.

For the class of Dunford–Pettis operators, N. Kalton and P. Saab [15] proved that if F is order continuous and T is Dunford–Pettis then R is also Dunford–Pettis.

1. Strictly singular operators

An operator $T \colon X \to Y$ between Banach spaces is said to be *strictly singular* (or *Kato*) if for every infinite dimensional (closed) subspace M of X, the restriction $T|_M$ is not an isomorphism into Y, i.e., there is not any infinite dimensional subspace M of X and $m > 0$ such that, for every $x \in M$,

$$m\|x\| \leq \|Tx\|.$$

This class forms a closed operator ideal, which properly contains the ideal of compact operators. For example, the inclusion operators $L^\infty[0,1] \hookrightarrow L^p[0,1]$, $1 \leq p < \infty$ are strictly singular (but no compact) (cf. [21, Theorem 5.2]). The strictly singular operator class is a very useful class in Fredholm and perturbation operator theory (cf. [2,16]).

It is well–known that an operator $T \colon X \to Y$ between Banach spaces is strictly singular if and only if for every infinite dimensional subspace M of X there exists another infinite dimensional subspace $N \subset M$ such that the restriction $T|_N$ is compact. In general, T (resp., the adjoint T^*) strictly singular does not imply that T^* (resp., T) is strictly singular.

A variant for Banach lattices is the following: given a Banach lattice E and a Banach space Y, an operator $T \colon E \to Y$ is called *disjointly strictly singular* if it is not invertible on any subspace of E generated by a disjoint sequence.

Clearly, every strictly singular operator is also disjointly strictly singular, but the converse is not true. For example, the canonical inclusions

$L^q[0,1] \hookrightarrow L^p[0,1]$, $1 \leq p < q < \infty$, are disjointly strictly singular but no strictly singular. This follows from the Khintchine inequality for the Rademacher functions (r_n):

$$\left(\int_0^1 |\sum a_n r_n|^p \, d\lambda\right)^{1/p} \sim \left(\sum_{n=1}^{\infty} |a_n|^2\right)^{1/2}.$$

In some special cases, both concepts of singularity coincide: for instance, for spaces with a Schauder basis of disjoint vectors or for $C(K)$-spaces. The class of all disjointly strictly singular operators is stable by addition and by composition by the right but in general it is not an operator ideal.

2. Strictly singular inclusions

Now we study the strict singularity and the disjoint strict singularity of inclusions between rearrangement invariant function spaces.

Recall that the *distribution function* λ_x of a measurable function x on $[0,\infty)$ is $\lambda_x(s) = \lambda\{t \geq 0 \colon |x(t)| > s\}$, and the *decreasing rearrangement* x^* of x is

$$x^*(t) = \inf\{s \geq 0 \colon \lambda_x(s) \leq t\}.$$

A Banach lattice E of measurable functions defined on $[0,\infty)$ is said to be a *rearrangement invariant space* (r.i. space) if $y \in E$ and $\lambda_x = \lambda_y$ imply $x \in E$ and $\|x\|_E = \|y\|_E$ (cf. [17]). The *fundamental function* ϕ_E of an r.i space E is defined by $\phi_E(t) = \|\chi_{[0,t]}\|_E$. Classical examples of r.i. spaces are Orlicz, Lorentz and Marcinkiewicz spaces. The *Orlicz space* L^φ consists of all measurable functions x on $[0,\infty)$ for which

$$\|x\|_{L^\varphi} = \inf\left\{s > 0 \colon \int_0^\infty \varphi\left(\frac{|x(t)|}{s}\right) dt \leq 1\right\} < \infty$$

where φ is a positive increasing convex function with $\varphi(0) = 0$.

Given E_1 and E_2 r.i. spaces, the sum space $E_1 + E_2$ with the norm

$$\|x\|_{E_1+E_2} = \inf\{\|x_1\|_{E_1} + \|x_2\|_{E_2} \colon x = x_1 + x_2, x_1 \in E_1, x_2 \in E_2\},$$

and the intersection space $E_1 \cap E_2$, with the norm $\|x\|_{E_1 \cap E_2} = \max(\|x\|_{E_1}, \|x\|_{E_2})$, are r.i. spaces. If $E_1 = L^1$ and $E_2 = L^\infty$ then

$$\|x\|_{L^1+L^\infty} = \sup_{\lambda(A)=1} \int_A |x(t)| \, dt = \int_0^1 x^*(t) \, dt.$$

Clearly $\phi_{L^1+L^\infty}(t) = \min(t,1)$, and the space coincides with the space of all locally integrable functions, i.e., $L^1 + L^\infty = L^1_{\text{loc}}(\lambda)$.

In the finite measure case it holds $L^\infty[0,1] \hookrightarrow E[0,1] \hookrightarrow L^1[0,1]$. The left canonical inclusions $L^\infty[0,1] \hookrightarrow E[0,1]$ are always strictly singular for any r.i. space $E \neq L^\infty$ (S. Novikov [19]). On the other side, for the right canonical inclusions $E[0,1] \hookrightarrow L^1[0,1]$, it holds that:

(i) It is disjointly strictly singular for any r.i. space $E \neq L^1$.
(ii) It is strictly singular if and only if $E[0,1]$ does not contain the order continous Orlicz space $L_0^{\exp x^2}[0,1]$.

In the statement (ii), proved by S. Novikov, E. Semenov and the author in [12], the necessity part follows from the Rodin–Semenov characterization of the r.i. spaces for which the Rademacher functions are equivalent to the canonical basis of ℓ^2 (cf. [17]).

In the infinite case $[0,\infty)$, we have the inclusions $L^1 \cap L^\infty \hookrightarrow E \hookrightarrow L^1 + L^\infty$. It turns out that the strict and the disjoint strict singularity of the left inclusions $L^1 \cap L^\infty \hookrightarrow E$ coincide and they are characterized in terms of the fundamental functions:

Theorem 2.1 ([13]). *Let E be an r.i. space. For the inclusion $L^1 \cap L^\infty \hookrightarrow E$, the following statements are equivalent: (i) strictly singular; (ii) disjointly strictly singular; (iii) weakly compact; (iv)* $\lim_{t\to 0} \phi_E(t) = \lim_{t\to\infty} \frac{\phi_E(t)}{t} = 0$.

Another strictly singular criterium has been obtained in [6] using interpolation.

The behavior of the right extreme inclusions $E \hookrightarrow L^1 + L^\infty$ is more diverse and tricky than the left inclusions, not coinciding none of the above three operator classes. Thus, conditions for the inclusion $E \hookrightarrow L^1 + L^\infty$ being strictly (resp., disjointly strictly) singular involve not only the behavior of the associated fundamental function. This arises from the non-disjoint strict singularity of the inclusions of the (order continuous) spaces $L_0^{p,\infty}$ into $L^1 + L^\infty$. Recall that $L^{p,\infty}$ consists of all measurable functions x on $[0,\infty)$ such that $\|x\|_{L^{p,\infty}} = \sup_{t>0} \left\{ t^{\frac{1}{p}} x^*(t) \right\} < \infty$.

Theorem 2.2 ([14]). *Let E be a r.i. space. The inclusion $E \hookrightarrow L^1 + L^\infty$ is strictly singular if and only if*

(i) $\lim_{t\to 0} \phi_E(t)/t = \lim_{t\to\infty} \phi_E(t) = \infty$.
(ii) $\sup_n \left\| t^{-\frac{1}{p}} \chi_{(1/n,n)} \right\|_E = \infty$ *for all* $1 < p < \infty$.
(iii) $E[0,1]$ *does not contain* $L_0^{\exp x^2}[0,1]$.

An important step in the proof of the above result is to show that conditions (i) and (ii) characterize precisely the *disjoint* strict singularity of the inclusion $E \hookrightarrow L^1 + L^\infty$. On the other hand, condition (i) characterizes the weak compactness.

3. Domination by strictly singular operators

We proceed to present domination results for positive strictly singular operators between Banach lattices.

First let us show that in general the domination property for this class is not true:

> *There exist operators $0 \leq R \leq T\colon \ell^1 \to L^\infty[0,1]$ such that T is strictly singular but R is not.*

Indeed, take $\widetilde{R}\colon \ell^1 \to L^\infty[0,1]$ the isometry defined by $\widetilde{R}(e_n) = r_n$. Consider also the positive operators $R_1, R_2\colon \ell^1 \to L^\infty[0,1]$ defined by $R_1(e_n) = r_n^+$ and $R_2(e_n) = r_n^-$ respectively, where r_n^+ and r_n^- are the positive and negative part of r_n. Clearly $\widetilde{R} = R_1 - R_2$. Moreover $0 \leq R_1, R_2 \leq T$, where T is the rank–one operator:

$$T(x) = \left(\sum_{n=1}^{\infty} x_n\right) \chi_{[0,1]}.$$

The operator T is strictly singular being compact, but neither the operator R_1 nor the operator R_2 are strictly singular. Now the equalities $T = R_1 + R_2$ and $\widetilde{R} = R_1 - R_2$ lead to contradiction.

We can give a general result using the above and the existence of ℓ^∞ sublattices in non order continuous Dedekind complete Banach lattices:

> *Let E and F be two Banach lattices with F Dedekind–complete; assume that neither E^* nor F are order continuous. Then there exist two positive operators $0 \leq R \leq T\colon E \to F$ such that T is strictly singular but R is not.*

A first step to present positive domination results is to consider the case when the range space of the operators is a $L^1(\mu)$-space, or more generally, an space with the positive Schur property. Recall that a Banach lattice E has the *positive Schur property* if every positive weakly null sequence is convergent. Examples of Banach lattices with the positive Schur property are the $L^1(\mu)$ spaces, the Orlicz spaces $L^{x \log^p(1+x)}[0,1]$ for $p > 0$, and the Lorentz spaces $L^{p,1}[0,1]$ for $1 < p < \infty$.

Observe that the positive Schur property implies that E does not contain an isomorphic copy of c_0 (in particular, E is order continuous). Otherwise, E would also contain a sequence of positive, pairwise disjoint elements $(e_n)_{n=1}^{\infty}$ equivalent to the unit vector basis of c_0 . This sequence must be weakly null and yet not convergent in norm, which gives a contradiction.

We also make use of the *Kadeč–Pełczyński disjointification* method for order continuous Banach lattices (cf. [17]): Let X be any subspace of an order continuous Banach lattice E. Then, either

(1) X contains an almost disjoint normalized sequence, that is, there exist a normalized sequence $(x_n)_{n=1}^{\infty} \subset X$ and a disjoint sequence $(z_n)_{n=1}^{\infty} \subset E$ such that $\|z_n - x_n\| \to 0$, or,
(2) X is isomorphic to a closed subspace of $L_1(\Omega, \Sigma, \mu)$.

Notice that if X is separable, then it can be included in some ideal H of E with a weak order unit. Therefore, this ideal has a representation as a Köthe function space over a finite measure space (Ω, Σ, μ) and, in this case, the previous dichotomy says that either X contains an almost disjoint sequence or the natural inclusion $J\colon H \hookrightarrow L_1(\Omega, \Sigma, \mu)$ is an isomorphism when restricted to X.

Let E and F be Banach lattices such that F has the positive Schur property. If $0 \le R \le T\colon E \to F$ and T is strictly singular, then R is strictly singular.

The proof of this result is based on factorizations of order weakly compact operators as well as properties of the class of M-weakly compact operators. Given a Banach lattice E and a Banach space Y, an operator $T\colon E \to Y$ is *order weakly compact* if $T[-x, x]$ is relatively weakly compact for every $x \in E_+$. And $T\colon E \to F$ is *M-weakly compact* if $\|Tx_n\| \to 0$ for every norm bounded disjoint sequence $(x_n)_n$ in E.

Order weakly compact operators can be characterized as those operators not preserving a positive disjoint order–bounded isomorphic copy of c_0 (cf. [18, Corollary 3.4.5]). Also, if X is a Banach space and F a Banach lattice, an operator $T\colon X \to F$ does not preserve an isomorphic copy of ℓ_1 complemented in F if and only if its adjoint T^* is order weakly compact (cf. [18]).

The following factorization result for positive order weakly compact operators is useful (see N. Ghoussoub and W. Johnson [11], also [5]):

Let E_1, E_2 be Banach lattices and operators $0 \le R \le T\colon E_1 \to E_2$. There exist a Banach lattice F, a lattice homomorphism $\phi\colon E_1 \to F$

and operators $0 \leq R^F \leq T^F$ *such that* $T = T^F\phi$ *and* $R = R^F\phi$*:*

$$\begin{array}{ccc} E_1 & \overset{T}{\underset{R}{\rightrightarrows}} & E_2 \\ & \phi\searrow \quad {}_{T^F}\nearrow\!\!\nearrow_{R^F} & \\ & F & \end{array}$$

And $T\colon E_1 \to E_2$ *is order weakly compact if and only if* F *is order continuous.*

A bounded subset A of a Banach lattice E is said to be *L-weakly compact* if $\|x_n\| \to 0$ for every disjoint sequence $(x_n)_n$ contained in the solid hull of A. If T is a regular operator from a Banach lattice E into a Banach lattice F with order continuous norm, then if A is a L-weakly compact subset of E we have $T(A)$ is also L-weakly compact.

If E is an order continuous Banach function space defined on a finite measure space (Ω, Σ, μ), a bounded subset $A \subset E$ is *equi-integrable* if for every $\varepsilon > 0$ there is $\delta > 0$ such that $\|f\chi_B\|_E < \varepsilon$ for every $B \in \Sigma$ with $\mu(B) < \delta$ and every $f \in A$.

If E is an order continuous lattice with a weak unit (hence, representable as an order ideal in $L^1(\Omega, \Sigma, \mu)$ for some probability space (Ω, Σ, μ)), then a bounded subset of E is equi-integrable if and only if it is L-weakly compact.

An order continuous Banach lattice E satisfies the *subsequence splitting property* (cf. [22]) if for every bounded sequence $(f_n)_n$ in E there is a subsequence $(n_k)_k$ and sequences $(g_k)_k$, $(h_k)_k$ in E with $|g_k| \wedge |h_k| = 0$ and

$$f_{n_k} = g_k + h_k$$

such that $(g_k)_k$ is equi-integrable and $|h_k| \wedge |h_l| = 0$ if $k \neq l$. Every p-concave Banach lattice ($p < \infty$) has the subsequence splitting property. Recall that E is p-concave if there exists a constant $M < \infty$ such that for every choice of elements $(x_i)_{i=1}^n$ we have

$$\left(\sum_{i=1}^n \|x_i\|^p\right)^{1/p} \leq M \left\|\left(\sum_{i=1}^n |x_i|^p\right)^{1/p}\right\|.$$

The following domination result for disjointly strictly singular operators given in [8] is also needed .

Let E *and* F *be Banach lattices such that* F *is order continuous. If* T *is disjointly strictly singular and* $0 \leq R \leq T\colon E \to F$ *then* R *is also disjointly strictly singular.*

We can now state a general domination result for strictly singular operators given recently by J. Flores, P. Tradacete and the author in [10], which improves a previous result given in [9] removing the order continuity hypothesis of E^*:

Theorem 3.1 ([10]). *Let E be a Banach lattice with the subsequence splitting property, and F an order continuous Banach lattice. If $0 \leq R \leq T: E \to F$ with T strictly singular, then R is strictly singular.*

In particular for the class of r.i. spaces we have:

> *Let E be a r.i. space that contains no isomorphic copy of c_0, F an order continuous Banach lattice and consider operators $0 \leq R \leq T: E \to F$ with T is strictly singular. Then R is strictly singular.*

Let us mention that applications to the strictly co-singular (or Pelczynski) operator class are given in [9].

4. Powers of dominated operators

In this section we study the so-called power operator problem for dominated endomorphisms.

When we consider Banach lattices $E = F$ and endomorphisms $0 \leq R \leq T: E \longrightarrow E$, it is interesting to study wether some iteration (power) of the operator R inherits a certain property of the operator T, under *no assumptions* on the Banach lattice E. This is called the *power problem* relative to a certain operator class.

This approach was first developed by C. D. Aliprantis and O. Burkinshaw in [3] and [4] for compact and weakly compact operators:

> *Let E be a Banach lattice and consider operators $0 \leq R \leq T: E \to E$.*
>
> *(i) If T is compact then R^3 is also compact.*
> *(ii) If T is weakly compact then R^2 is also weakly compact.*

The power problem for the class of Dunford–Pettis operators was studied by N. Kalton and P. Saab in [15] and, for the class of disjointly strictly singular operators, by J. Flores and the author in [8]:

> *Let E be a Banach lattice and consider operators $0 \leq R \leq T: E \to E$.*
>
> *(i) If T is Dunford–Pettis then R^2 is also Dunford-Pettis.*

(ii) If T is disjointly strictly singular then R^2 is also disjointly strictly singular.

All these results are optimal in the sense that it is possible to produce counterexamples when the powers of R are lower.

First let us show that the power problem for strictly singular endomorphisms is not trivial:

There exist operators $0 \leq R \leq T\colon L^2[0,1] \oplus \ell^\infty \to L^2[0,1] \oplus \ell^\infty$ such that T is strictly singular but R is not.

Indeed, consider the rank–one operator $Q\colon L^1[0,1] \to \ell_\infty$ defined by

$$Q(f) = \left(\int_0^1 f, \int_0^1 f, \ldots \right).$$

Take also an isometry $S\colon L^1[0,1] \to \ell^\infty$ given by $S(f) = (h'_n(f))_{n=1}^\infty$, where $(h_n)_{n=1}^\infty$ is a dense sequence in the unit ball of $L^1[0,1]$, and $(h'_n)_{n=1}^\infty$ is a sequence of norm one functionals such that $h'_n(h_n) = \|h_n\|$ for all n. If $J\colon L^2[0,1] \hookrightarrow L^1[0,1]$ denotes the canonical inclusion, then the operator $SJ\colon L^2[0,1] \to \ell_\infty$ is not strictly singular.

Since ℓ^∞ is Dedekind complete we have that $|SJ|$, $(SJ)^+$ and $(SJ)^-$ are also continuous operators between $L^2[0,1]$ and ℓ^∞. It is easy to see that $|SJ| \leq QJ$. Since SJ is not strictly singular, we must have that either $(SJ)^+$ or $(SJ)^-$ is not strictly singular, so let us assume, w.l.o.g., that $(SJ)^+$ is not strictly singular. Now consider the matrices of operators:

$$R = \begin{pmatrix} 0 & 0 \\ (SJ)^+ & 0 \end{pmatrix}, \qquad T = \begin{pmatrix} 0 & 0 \\ QJ & 0 \end{pmatrix},$$

which clearly define operators with the required properties.

Next, we show some positive results:

Theorem 4.1. *Let E be a Banach lattice and consider operators $0 \leq R \leq T\colon E \to E$. If T is strictly singular, then R^4 is also strictly singular.*

This is deduced from a more general statement proved by J. Flores, P. Tradacete and the author in [10] using factorization methods:

Theorem 4.2. *Let*

$$E_1 \underset{R_1}{\overset{T_1}{\rightrightarrows}} E_2 \underset{R_2}{\overset{T_2}{\rightrightarrows}} E_3 \underset{R_3}{\overset{T_3}{\rightrightarrows}} E_4 \underset{R_4}{\overset{T_4}{\rightrightarrows}} E_5$$

be operators between Banach lattices, such that $0 \leq R_i \leq T_i$ *for* $i = 1, 2, 3, 4$. *If* T_1, T_3 *are strictly singular and* T_2, T_4 *are order weakly compact then* $R_4 R_3 R_2 R_1$ *is also strictly singular.*

As a direct consequence we have Theorem 4.1. Indeed, since T is strictly singular, it cannot preserve an isomorphic copy of c_0 so, in particular, it is order weakly compact. Therefore, it suffices to apply Theorem 4.2 to $E_i = E$, $R_i = R$ and $T_i = T$ for all i.

Corollary 4.1. *Let* $0 \leq R \leq T \colon E \to F$ *and* $0 \leq S \leq V \colon F \to G$. *If* F *and* G *are order continuous Banach lattices, and* T *and* V *are strictly singular operators, then the composition* SR *is strictly singular.*

In particular, if $0 \leq R \leq T \colon E \to E$ *with* T *strictly singular and* E *order continuous, then* R^2 *is strictly singular.*

Indeed, since F is order continuous, the identity $I_F \colon F \to F$ is order weakly compact. Consider $E_1 = E$, $E_2 = F$, $E_3 = F$, $E_4 = G$ and $E_5 = G$; and the operators $T_1 = T$, $T_2 = I_F$, $T_3 = V$ and $T_4 = I_G$. Then, by Theorem 4.2, we obtain that $I_G S I_F R = SR$ is strictly singular.

Note that in the above example the lattice $L^2[0,1] \oplus \ell^\infty$ is not order continuous and the square R^2 is the zero operator (hence, strictly singular).

A open *question* is the following: Do there exist an order continuous Banach lattice E and operators $0 \leq R \leq T \colon E \to E$ such that T is strictly singular but R is not?

Acknowledgements: I would like to thank the organizers of the III Course of Mathematical Analysis at La Rabida (Huelva) for the invitation and the kind hospitality.

References

1. Y. A. Abramovich and C. D. Aliprantis, *An invitation to operator theory*, Graduate Studies in Mathematics, 50. American Mathematical Society, Providence, RI (2002).
2. Y. A. Abramovich and C. D. Aliprantis, *Positive operators*, In Handbook of the Geometry of Banach spaces, vol I. Edited by W. Johnson and J. Lindenstrauss. Elsevier (2001).
3. C. D. Aliprantis and O. Burkinshaw, *Positive compact operators on Banach lattices*, Math. Z. **174** (1980), 289–298.
4. C. D. Aliprantis and O. Burkinshaw, *On weakly compact operators on Banach lattices*, Proc. Amer. Math. Soc. **83** (1981), 573–578.
5. C. D. Aliprantis and O. Burkinshaw, *Positive Operators*, Academic Press (1985).

6. F. Cobos, A. Manzano, A. Martinez and P. Matos, *On interpolation of strictly singular operators, strictly co-singular operators and related operator ideals*, Proc. Roy. Soc. Edinburgh **130A** (2000) 971–989.
7. P. G. Dodds and D. H. Fremlin, *Compact operators in Banach lattices*, Israel J. Math. **34** (1979), 287–320.
8. J. Flores and F. L. Hernández, *Domination by positive disjointly strictly singular operators*, Proc. Amer. Math. Soc. **129** (2001), 1979–1986.
9. J. Flores and F. L. Hernández, *Domination by positive strictly singular operators*, J. London Math. Soc. **66** (2002), 433–452.
10. J. Flores, F. L. Hernández and P. Tradacete, *Powers of operators dominated by strictly singular operators*, Quart. J. Math. Oxford (to appear).
11. N. Ghoussoub and W. B. Johnson, *Factoring operators through Banach lattices not containing $C(0,1)$*, Math. Z. **194** (1987), 153–171.
12. F. L. Hernández, S. Novikov and E. M. Semenov, *Strictly singular embeddings between rearrangement invariant spaces*, Positivity **7** (2003), 119–124.
13. F. L. Hernández, V. M. Sánchez and E. M. Semenov, *Disjoint strict singularity of inclusions between rearrangement invariant spaces*, Studia Math. **144** (2001), 209–226.
14. F. L. Hernández, V. M. Sánchez and E. M. Semenov, *Strictly singular inclusions into $L^1 + L^\infty$*, Math. Z. **258** (2008), 87–106.
15. N. Kalton and P. Saab, *Ideal properties of regular operators between Banach lattices*, Illinois J. Math **29** (1985), 382–400.
16. J. Lindenstrauss and L. Tzafriri, *Classical Banach Spaces I: Sequence Spaces*, Springer–Verlag, Berlin (1977).
17. J. Lindenstrauss and L. Tzafriri, *Classical Banach Spaces II: Function Spaces*, Springer–Verlag, Berlin (1979).
18. P. Meyer–Nieberg, *Banach Lattices*, Springer–Verlag (1991).
19. S. Novikov, *Singularities of embedding operators between symmetric function spaces on $[0,1]$*, Math. Notes **62** (1997), 457–468.
20. L. Pitt, *A compactness condition for linear operators on function spaces*, J. Operator Theory **1** (1979), 49–54.
21. W. Rudin, *Functional Analysis*, McGraw-Hill, New York (1977).
22. L. Weis, *Banach lattices with the subsequence splitting property*, Proc. Amer. Math. Soc. **105**(1) (1989), 87–96.
23. A. W. Wickstead, *Extremal structure of cones of operators*, Quart. J. Math. Oxford Ser. (2) **32** (1981), 239–253.

THE HAHN–BANACH THEOREM AND THE SAD LIFE OF E. HELLY

LAWRENCE NARICI[†] and EDWARD BECKNSTEIN[‡]

Mathematics Department
St. John's University
Staten Island, NY 11301, USA
E-mails: [†]*narici@gmail.com,* [‡]*drbeckense@aol.com*

1. Introduction

Let M be a linear subspace of a normed space X and let $f : M \to \mathbf{K} = \mathbf{R}$ or $\mathbf{C}$ be a continuous linear functional. An analytic form of the Hahn–Banach theorem asserts the existence of a continuous linear extension of f to X. Its consequences ring throughout functional analysis and other disciplines. The Austrian mathematician Eduard Helly (1884-1943) played a significant role played in the development of the theorem and we outline here what he did and how he did it. We also discuss Helly's life, from his poorly tended wound in the first world war, to his flight from the Nazis, to his early death. Although he only published five papers in journals — and only two of those in functional analysis proper — his largely overlooked contributions are quite significant. In the context of complex sequence spaces he defined an abstract normed space X, a seminormed dual X^d (Sec. 4) and proved [1921] a Hahn–Banach theorem for real or complex normed sequence spaces X for which X^d was separable. The key to his argument was the one-dimensional extension, the ability to continuously extend f to $M \oplus \mathbf{K}x$ for $x \notin M$. Hahn and Banach both used the one-dimensional extension technique, but accomplished it by different means. Moreover, although they did not require a separable dual, they only proved it for real spaces. Helly's approach was more geometric. He used the following "intersection theorem", a result that he discovered but which was first published by Johann Radon [1921]:

Theorem 1.1 (Helly's Intersection Theorem I (Helly [1923])). *Let $C_1, C_2, \ldots, C_m$ be a finite number of convex subsets of $\mathbf{R}^n$ with $m > n$.*

If every $n+1$ of the C_i meet, then their intersection is nonempty.

Ultimately [1930], Helly generalized it to:

Theorem 1.2 (Helly's Intersection Theorem II). *If any $n+1$ members of an arbitrary family of* compact *convex subsets of $\mathbf{R}^n$ meet, then their intersection is nonvoid.*

Things came full circle in the 1970's and early 1980's. Various authors used intersection properties (Sec. 7) like these to avoid the standard reduction–to–the–real–case argument to deduce the complex Hahn–Banach theorem from the real one. You can prove it (Sec. 7) pretty much simultaneously when the underlying field is $\mathbf{R}$, $\mathbf{C}$, the quaternions, or certain fields ("spherically complete") with a non-Archimedean valuation. Intersection properties also characterize the normed spaces Y which may be substituted for the field $\mathbf{K}$ in the Hahn–Banach theorem (Sec. 7, Theorem 7.1). Other important applications of Helly's intersection theorems appear in areas as distant from functional analysis as the study of DNA molecules (see Grünbaum and Klee [1967] for a discussion of this and others).

2. The origin of the Hahn–Banach Theorem

For there to be a solution to a finite system of linear equations, the equations have to be "compatible", in that they cannot require contradictory things. To determine compatibility for infinite systems, the first attempts extended known techniques. Basically, it was classical analysis — almost solve the problem in some kind of finite situation, then take a limit. A fatal defect in this case was the need for the (very rare) convergence of infinite products.

Two problems of particular interest in the late nineteenth century which directly led not just to the Hahn–Banach theorem but to the invention of normed spaces in general were the moment and Fourier series problems:

- **The moment problem.** If all moments $f_n(x) = \int_0^1 t^n x(t)\, dt = c_n$ ($n \in \mathbf{N}$) of a function x are known, find x.
- **The Fourier series problem.** If all Fourier coefficients of a function x are known, find x.

Riesz and Helly obtained solutions to problems like these in important special cases such as $L_p\,[0,1]$ and $C\,[a,b]$. In modern language, they discovered that compatibility was equivalent to the continuity of a certain linear functional. Consider more general versions of the problems above: Let X

be a normed space with dual X', let S be a set, and let $\{c_s : s \in S\}$ be a collection of scalars.

(V) **The vector problem.** Let $\{f_s : s \in S\}$ be a collection of bounded linear functionals on X. Find $x \in X$ such that $f_s(x) = c_s$ for every s. and its dual:

(F) **The functional problem.** Let $\{x_s : s \in S\}$ be a collection of vectors from X. Find $f \in X'$ such that $f(x_s) = c_s$ for every s.

If X is reflexive then solving (F) also solves (V), for given "vectors" $\{f_s : s \in S\} \subset X'$ there exists $h \in X''$ such that $h(f_s) = c_s$ for every s. Now choose the $x \in X$ such that $h(f_s) = f_s(x)$ for every s.

Motivated by Hilbert's work on $L_2[0,1]$, Riesz [1910] invented the spaces $L_p[0,1]$, $1 < p < \infty$ (he didn't consider the ℓ_p spaces until 1913). Instead of the moment and Fourier series problems *per se* [1910, 1911], he considered the vector problem (LP) below. In doing so, he inadvertently proved a special case of the Hahn–Banach theorem.

(LP) Let S be a set. For $p > 1$ and $1/p + 1/q = 1$, given y_s in $L_q[a,b]$ (equivalently, consider the functionals f_s of Eq. (1)) and scalars $\{c_s : s \in S\}$, find $x \in L_p[a,b]$ such that

$$f_s(x) = \int_a^b x(t) y_s(t) dt = c_s \text{ for each } s \in S \tag{1}$$

For there to be such an x, he showed that the following necessary and sufficient connection between the y's and the c's had to prevail: There exists $K > 0$ such that for any finite set of indices s and scalars a_s,

$$\left|\sum a_s c_s\right| \leq K \left(\int_a^b \left|\sum a_s y_s\right|^q \right)^{1/q} = K \left\|\sum a_s y_s\right\|_q \tag{*}$$

Condition (*) implies that if the y's are linearly dependent, i.e., $\sum a_s y_s = 0$ for a finite set of scalars a_s, then $\sum a_s c_s = 0$ as well. Thus, if we consider the linear functional g on the linear span $M = [y_s : s \in S]$ of the y's in $L_q[a,b]$ defined by taking $g(y_s) = c_s \ (s \in S)$, g is well-defined. Not only that, for any y in M, $|g(y)| \leq K \|y\|_q$ on M, so g is continuous on M. If there is an x in L_p which solves (LP), then g has a continuous extension G to L_q, namely, for any y in L_q,

$$G(y) = \int x(t) y(t) \, dt \quad [G(y_s) = c_s \ (s \in S)]$$

Thus, Riesz showed that:

- (LP) is solvable if and only if a certain linear functional g defined on a subspace of L_q is continuous.
- If the system is solvable, then g can be extended to a continuous linear functional defined on all of L_q.

3. Helly

Eduard Helly was born in Vienna in 1884 and got his Ph. D. from the University of Vienna in 1907 (a reproduction of the first page of his *handwritten* dissertation appears on p. 130 of Butzer *et al.* [1980]; that article and another by Butzer and others in 1984 are excellent sources for information about Helly and his work). By means different from and simpler than Riesz [1911], Helly also solved a moment problem in 1912 and proved special cases of the Hahn–Banach and Banach–Steinhaus theorems for linear functionals on $C[a,b]$.

Helly volunteered for the Austrian Army in 1914 and went to the Russian front in 1915. He was wounded by a bullet through the lungs in September 1915, a wound that ultimately caused his death. He spent almost the next five years as a prisoner of war in a camp near Tobolsk, Siberia. He endured eastern Siberia's frigidity along with a Hungarian university student named Tibor Radó (1895-1965). Helly tutored Radó in the camp and imbued him with a taste for mathematical research. Ultimately, Radó became a distinguished mathematician. The Great War ended but peace did not come to Russia. The White Russian forces contended with the Red armies. Other players on the scene were a Czech army of some 50,000 escaped prisoners who joined the White Russians. Japan saw an opportunity to pick up some Russian territory and sent troops; so did the Americans, the British and others. In the midst of this chaos, there was no repatriation of POWs. Radó escaped from the camp in 1919 and went north! With the help of some Eskimos, he eventually traversed thousands of kilometers on his way west and reached Hungary in 1920. He abandoned civil engineering and switched to mathematics at the University of Szeged. His teachers included Frigyes Riesz and Alfred Haar. His most famous work is his solution to the Plateau problem published in 1930 concerning bounding contours for minimal surfaces.

As of the summer of 1920, Helly was still a POW in Tobolsk but by first going east to Japan, then to the Middle East and Egypt, Helly got back to Vienna in mid-November of 1920. In order to be a professor in the Austrian system, it was necessary to write a post-doctoral thesis called a *Habilitationsschrift.* As with a doctoral dissertation, it is reviewed by and defended before an academic committee. It is necessary to attain the Habilitation (the qualification) to be a *Privatdozent*, one who may supervise doctoral students. Helly had presented talks about what became his 1921 paper to the Viennese Mathematical Association (Wiener Mathematischen Gesellschaft) before the war. He successfully presented his Habilitation thesis to the faculty of the University of Vienna in 1921 and then applied for a professorship there. Largely as a result of Hahn's opposition, he did not get one. Helly's wife, Dr. Elise Bloch, also a mathematician, attributed Hahn's opposition to two sources: Helly was (1) Jewish and (2) too old, Helly being 37 at the time. Helly did become a Privatdozent in August 1921, a position that paid nothing. To support himself, he went to work for a bank. As a Privatdozent, he supervised three doctoral students and taught practically every semester from 1921 until 1938. The bank failed in 1929 and he got a job in 1930 in the actuarial department of an insurance company, Lebensversicherungs–Gesellschaft Phönix where his co-workers included the mathematicians Eugene Lukacs (who had taken courses with Helly and Hahn at the University of Vienna) and Z. W. Birnbaum. In a 1979 letter (Butzer *et al.* [1980, p. 139]), Birnbaum said:

> Helly was a delightful man, cheerful in the face of adversities, with a gentle sense of humor. There were three mathematicians in the Phönix office who were my immediate superiors. One of them had the title "Prokurist" while Helly, to my knowledge, did not get that high. Whenever a non-routine question came up, the difference between Helly and the other two became apparent: Helly gave the problem a mathematical formulation and obtained a solution which could be used over and over again in similar cases; the other two worked the problem numerically in each case, by trial and error, grinding it out on their hand-operated Odhner desk calculators. Incidentally, even the manner in which he handled his desk calculator was ingenious, devising shortcuts and step–saving routines.

On March 13, 1938, the day after the *Anschluss Österreichs*, the political union of Germany and Austria, Jews were ordered to appear in evening dress and scrub the streets. Stores and apartments were pillaged. In May 1938 the Nazis enacted the Nuremberg racial laws. These excluded Jews from most professions, barred them from attending universities and forced them to wear a yellow badge. All Jewish women had to take the name Sarah as part of their name, all Jewish men the name Israel. All Jewish bank accounts were frozen and all licenses held by Jews—even driver's licenses—were revoked. As a result, Helly was fired by the Phönix and could no longer teach at the university. It was still possible, indeed encouraged, for Jews to emigrate after paying an emigration tax, the *Reichsfluchtsteuer* and 130,000 did. Among the 30,000 who came to the US were Helly, his wife, and their eight-year old son Walter Sigmund; they emigrated to Brooklyn in 1938 (Birnbaum and Lukacs also emigrated to the US). There were so many qualified *émigrés* that Helly was unable to secure a university position, even though he had letters of recommendation from Einstein, Oswald Veblen and Hermann Weyl. He survived by tutoring high school students. Eventually, he found employment at some junior colleges in New Jersey. In 1943, upon the recommendation of Karl Menger among others, he was offered the position of visiting lecturer at Illinois Institute of Technology in Chicago. This turn of good fortune did him no good, however. His second heart attack, a remnant of his WWI wound, killed him on November 28, 1943. Gödel, a thesis student of Hahn's, summed it up in a note to Walter in the funeral book: "Now all is well but . . .Papa dies". Walter got a Ph. D. in physics from the Massachusetts Institute of Technology and later became Professor of Operations Research at the Polytechnic Institute of Brooklyn, our *alma mater*, now called Polytechnic University. He is noted for "(p, q)-Helly cliques" and also his 1975 book "Urban Systems Models". We were at "Poly" as students and teachers until 1967 but never met him. A friend of ours, Maurice Figueres, took a course with him there in 1987 and was frequently driven back to Manhattan by him after class. Maurice quoted him as saying "My dad was a *real* mathematician".

4. The "Landmark": Helly's 1921 article

Helly published the results of his Habilitationsschrift in 1921 in an article. Dieudonné [1981, p. 130] subsequently called "a landmark in the history of functional analysis". As Helly says at the beginning of the article, the conditions for solving infinite systems of linear equations had been given by Schmidt [1908] and Riesz [1913] "in the case that the coefficients and solutions satisfy certain inequalities". His aim, he said, was to show that the conditions could be interpreted geometrically. Some high points of the article are:

★ **General normed sequence space.** He abandoned special cases and defined a general normed sequence space $X \subset \mathbf{C}^{\mathbf{N}}$, although he did not require that X be a vector space. He assumes that there is a norm D defined on X; he did not use the word *norm*, or the notation $\|\cdot\|$.

★ **Dual space.** Helly took as the "dual space" of X the set X^d of all complex sequences $u = (u_n)$ such that $\sum_{n\in\mathbf{N}} x_n u_n < \infty$ for all $(x_n) \in X$. He did not give X^d a name. He did call the seminorm Δ (see below) on X^d a *polare Abstandsfunktion*, so he may have been thinking of *polare Raum*, the name Hahn later used for the dual in a more general setting. X^d is a vector space regardless of what X is. If $X = c$ or c_0, then $X^d = \ell_1$; if $X = \ell_1$, then $X^d = \ell_\infty$ but if $X = \ell_\infty$, the X^d you get is a proper subset of what we call the dual X' of X today. Nowadays, such pairs (X, X^d), subject to *absolute* convergence of $\sum x_n u_n$, are called *Köthe sequence spaces* and *α-duals*, respectively.

★ **Seminorm for the dual.** For $x = (x_n) \in X$ and $u = (u_n) \in X^d$, Helly defines an analog of an inner product $\langle\cdot,\cdot\rangle$ on $X \times X^d$: for x in X and u in X^d, $\langle x, u\rangle = \sum_{n\in\mathbf{N}} x_n u_n$ (if X is a vector space, $\langle\cdot,\cdot\rangle$ is a bilinear form and (X, X^d) a dual pair). Using an idea of Minkowski's, he defines the *polare Abstandsfunktion* Δ for X^d as

$$\Delta(u) = \sup\{|\langle x, u\rangle| : D(x) = 1\}.$$

He observes that Δ is generally a seminorm, not a norm. In the event that Δ is not a norm, he notes that each point of X is in some subspace Y of $\mathbf{C}^{\mathbf{N}}$ of codimension 1 (to see this, suppose $u = (u_n) \neq 0$ and, in particular, $u_1 \neq 0$; if there exists a nonzero unit vector $x = (c, 0, 0, \dots) \in X$ then $|\langle x, u\rangle| = |cu_1| \neq 0$ which implies that $\Delta(u) \neq 0$). He notes that D and Δ satisfy a Cauchy–Schwarz–type inequality, namely,

$$|\langle x, u\rangle| \leq D(x)\,\Delta(u).$$

★ **The problem.** Helly sought to solve the following vector problem:

- Given sequences $f_n = (f_{nj})$ from $X^d \subset \mathbf{C}^{\mathbf{N}}$ and a sequence $(c_n) \in \mathbf{C}^{\mathbf{N}}$, find $x = (x_j) \in X$ such that

$$\langle x, f_n\rangle = \sum_{j\in\mathbf{N}} x_j f_{nj} = c_n \text{ for each } n \in \mathbf{N}.$$

His method of attack is quite original: he doesn't seek the x right away. Apparently with a belief in reflexivity (when he began his investigation) he seeks:

(1) a continuous linear functional $h \in X^{dd} = (X^d)^d$ such that $h(f_n) = c_n$ for each n, then
(2) $x \in X$ such that $h(f_n) = f_n(x)$ for every n.

He discovered that the $x \in X$ corresponding to h did not always exist, thus showing that some spaces are not reflexive.

The first Hahn–Banach theorem. In order to establish (1), Helly extended a bounded linear functional f from a subspace M to the whole space. *Assuming that X^d is separable* so he could use induction, the key step was the one-dimensional extension: For x not in M, find a linear functional F such that, for $\mathbf{K} = \mathbf{R}$ or $\mathbf{C}$,

$$\begin{array}{lccl} F: M \oplus \mathbf{K}x & & & |F| \leq k\,\|\cdot\| \\ \qquad \Big| & \searrow & & \\ f: \qquad M & \longrightarrow & \mathbf{K} & |f| \leq k\,\|\cdot\| \text{ (for some } k) \end{array}$$

Hahn [1927] and Banach [1929] used the technique of the one-dimensional extension as well to prove what we call the Hahn–Banach theorem today. As they used transfinite induction rather than ordinary induction, they generalized it by eliminating the separability of the dual. Each acknowledged Helly's work. Their gain in generality was offset by the fact that their proof required the choice of a number between two others and so only applied to real spaces. Helly used certain intersection properties that we discuss in Sec. 5.

In one of the first important applications of the Hahn–Banach theorem, Banach [1932, pp. 55–57, Theorems 4 and 5] solved the general functional problem. In the proof of sufficiency in Theorem 4.1 below, he used condition (**) to create a continuous linear functional on a subspace which he then extended to the whole space by the Hahn–Banach theorem. As he was generalizing a result of Helly [1912] (and Riesz [1910a]), Theorem 4.1 is usually referred to as *Helly's theorem.*

Theorem 4.1. *Let X be a real normed space, let $\{x_s\}$ and $\{c_s\}$, $s \in S$, be sets of vectors and scalars, respectively. Then there is a continuous linear functional f on X such that $f(x_s) = c_s$ for each $s \in S$ if and only if there exists $K > 0$ such that for all finite subsets $\{s_1, \ldots, s_n\}$ of S and scalars $a_1, \ldots, a_n$*

$$\left| \sum_{i=1}^{n} a_i c_{s_i} \right| \leq K \left\| \sum_{i=1}^{n} a_i x_{s_i} \right\| \qquad (**)$$

5. Helly's technique

Helly had observed in his 1912 paper that *any* collection of mutually intersecting closed intervals $\{[a_s, b_s] : s \in S\}$ has a nonempty intersection (Proof: no left endpoint a_s can be greater than any right endpoint b_t). He later extended this to *finite* collections of convex subsets of $\mathbf{R}^n$, namely, that if any $n+1$ members of a family of $m > n$ convex subsets of $\mathbf{R}^n$ meet, then their intersection is nonvoid (to see the need for a *finite* number of sets, consider the collection of closed upper half-planes in $\mathbf{R}^2$). Helly had lectured on the result about convex sets to the Viennese Mathematical Union in 1913, but did not publish it until 1923. In the meantime, Radon published the result in 1921. To extend a certain continuous linear functional from a subspace to the whole space, Helly observed that certain inequalities had to be satisfied; he further observed that satisfaction of these inequalities was equivalent to the nonemptiness of an intersection of a finite collection of disks in $\mathbf{C}$. By using his intersection theorem (Theorem 1.1 of the Introduction), he reduced the problem to showing that any three of these disks had nonempty intersection. In the process, he deduced a Hahn–Banach theorem for real or complex normed sequence spaces X when $\left(X^d, \Delta\right)$ was separable.

6. The complex case

Although Helly proved his Hahn–Banach theorem for certain complex sequence spaces, the complex version for the general case languished until 1936. The key is the intimate relationship between the real and complex parts of a complex linear functional f, namely that

$$\operatorname{Re} f(ix) = \operatorname{Im} f(x).$$

Although usually credited to F. Murray [1936], H. Löwig discovered this in 1934. Murray reduced the complex case to the real case, then used the real Hahn–Banach theorem to prove the complex form for subspaces of $L_p[a, b]$ for $p > 1$. Murray's perfectly general method was used and acknowledged by Bohnenblust and Sobczyk [1938] who proved it for arbitrary complex normed spaces. They, incidentally, were the first to call it the Hahn–Banach theorem. Also by reduction to the real case, Soukhomlinov [1938] and Ono [1953] obtained the theorem for normed spaces over the complex numbers and the quaternions.

We consider the complex case next in a way that does not depend on reducing the complex case to the real case.

7. Intersection properties

Intersection properties can not only be used to prove the Hahn–Banach theorem for normed spaces over $\mathbf{R}$, $\mathbf{C}$, the quaternions $\mathbf{H}$, or certain fields F with a non-Archimedean valuation. They are also useful in solving:

The extension problem. What normed spaces Y can replace the scalar field $\mathbf{K}$ in the Hahn–Banach theorem? Specifically, if $A : M \to Y$ is a bounded linear map on the closed subspace M of the normed space X, under what circumstances is there is a linear extension $\bar{A} : X \to Y$ of A such that $\|\bar{A}\| = \|A\|$?

If such an $\bar{A}$ exists for any A on any subspace M of any normed space X, we say that Y is **extendible.** Banach and Mazur showed that there are non-extendible spaces in 1933. An obvious difficulty of characterizing extendible spaces Y is that the A, the M and the X have nothing to do with Y. Nevertheless, Nachbin [1950] internally characterized extendible *real* normed spaces Y as those for which any family of mutually intersecting closed balls has nonempty intersection. But back to the problem for linear functionals.

The key intersection property. Let X denote a normed space over a field $\mathbf{K}$ with an absolute value $|\cdot|$. For $r > 0$, $B(0, r) = \{c \in \mathbf{K} : |c| \leq r\}$. Let f denote a continuous linear functional defined on a subspace M of X. As the key to the proof of the Hahn–Banach theorem is the one-dimensional extension, we need to know: For what $\mathbf{K}$ can we extend f from M to a continuous linear functional F defined on $M \oplus \mathbf{K}x$ for any $x \notin M$ with $\|F\| = \|f\|$? We may clearly assume that $\|f\| \leq 1$ and, since we can extend f by continuity to the closure of M, that M is closed.

To preserve the bound, it is necessary and sufficient to find a value a for $F(x)$ that satisfies $|F(x) - f(m)| = |a - f(m)| \leq \|x - m\|$ for all $m \in M$. In other words, a must lie in $B(f(m), \|x - m\|)$ for every $m \in M$, i.e., must belong to $\bigcap_{m \in M} B(f(m), \|x - m\|)$. To extend f, it is therefore necessary and sufficient that $\mathbf{K}$ satisfy the following intersection property:

$$\bigcap_{m \in M} B(f(m), \|x - m\|) \neq \emptyset \qquad \text{(HBIP)}$$

The following intersection properties are equivalent to (HBIP) but they are purely internal in that they do not involve the $x \in X$ or the subspace M of X.

Definition 7.1 (Intersection Properties). *Let S be a set, $\{c_s : s \in S\}$ a collection of scalars and $\mathcal{B} = \{B(c_s, r_s) : c_s \in \mathbf{K},\ r_s > 0,\ s \in S\}$ a collection of closed balls in $\mathbf{K}$. If $\bigcap \mathcal{B} \neq \emptyset$ whenever:*

*(a) $\bigcap \{B(bc_s, r_s) : s \in S\} \neq \emptyset$ (in $\mathbf{K}$) for any $|b| \leq 1$, then $\mathbf{K}$ has **the weak intersection property;***

*(b) for any finite subcollection $B(c_{s_k}, r_{s_k}) \in \mathcal{B}$, $k = 1, 2, \ldots, n$, $n \in \mathbf{N}$, and any scalars $b_1, b_2, \ldots, b_n \in \mathbf{K}$, $\sum_{k=1}^{n} b_k = 0$ implies that $\|\sum_{k=1}^{n} b_k c_{s_k}\| \leq \sum_{k=1}^{n} |b_k| r_{s_k}$, then $\mathbf{K}$ has **Holbrook's intersection property.***

Let $\mathbf{K} = \mathbf{R}$, $\mathbf{C}$ or $\mathbf{H}$, the quaternions. Since the closed balls $B(bc_s, r_s)$ of $\mathbf{K}$ are convex and compact, $\bigcap \mathcal{B} \neq \emptyset$ if any two, three or five, respectively, of the $B(c_s, r_s)$ have nonempty intersection by Helly's second intersection theorem — in particular, the "finite" in Holbrook's intersection property can be replaced by 2, 3, or 5 in those cases. Instead of reduction to the real case, Hustad [1973], Holbrook [1975] and Mira [1982] used these properties to prove the Hahn–Banach theorem for $\mathbf{R}$, $\mathbf{C}$ or $\mathbf{H}$ simultaneously with no reduction to the real case. Mira corrected an error in Holbrook's argument and also showed that if $\mathbf{K}$ is a non-Archimedean valued field (assuming that X has a norm which also satisfies the ultrametric triangle inequality) then we need only require nonempty intersections for each pair of elements of $\mathcal{B}$.

As to the general extension problem, consider the following:

Definition 7.2 (Intersection Properties). *Let S be a set and let $\mathcal{B} = \{B(y_s, r_s) : y_s \in Y,\ r_s > 0,\ s \in S\}$ be a collection of closed balls in the normed space Y with dual Y'. If $\bigcap \mathcal{B} \neq \emptyset$ whenever:*

(a) $\bigcap \{B(f(y_s), r_s) : s \in S\} \neq \emptyset$ *(in $\mathbf{K}$) for each f in the unit ball of Y', then Y has the* ***weak intersection property;***

(b) for any $B(y_{s_k}, r_{s_k}) \in \mathcal{B}$, $k = 1, 2, \ldots, n$, $n \in \mathbf{N}$, and $b_1, b_2, \ldots, b_n \in \mathbf{K}$, $\sum_{k=1}^{n} b_k = 0$ implies that $\|\sum_{k=1}^{n} b_k y_{s_k}\| \leq \sum_{k=1}^{n} |b_k| r_{s_k}$, then Y has ***Holbrook's intersection property.***

(V) **The vector problem.** Let $\{f_s : s \in S\}$ be a collection of bounded linear functionals on X. Find $x \in X$ such that $f_s(x) = c_s$ for every s. and its dual:

(F) **The functional problem.** Let $\{x_s : s \in S\}$ be a collection of vectors from X. Find $f \in X'$ such that $f(x_s) = c_s$ for every s.

Theorem 7.1 (Extendible Spaces). *A Banach space Y over $\mathbf{K} = \mathbf{R}$, or $\mathbf{C}$ is extendible if and only if any of the three equivalent properties below is satisfied:*

(HBIP) *For any subspace M of any normed space X, any continuous linear map $A\colon M \to Y$ and any $x \notin M$, $\bigcap_{m \in M} B(Am, \|x - m\|) \neq \emptyset$.*

(WIP) [Hustad 1973] *Y has the weak intersection property.*

(HIP) [Holbrook 1975, Mira 1982] *Y satisfies Holbrook's intersection property.*

Holbrook's intersection property is concerned with inequalities about centers and radii of certain balls. It is not obvious that it has anything to do with intersections. It is straightforward to show that two closed balls $B(x,r)$ and $B(y,s)$ in a normed space meet if and only if the distance $\|x-y\|$ between their centers is less than or equal to the sum $r+s$ of their radii: $\|x-y\| \leq r+s$. Suppose $\mathcal{B}$ is a collection of closed balls (closed intervals) in $\mathbf{R}$ satisfying Holbrook's intersection property and $B(x,r)$, $B(y,s) \in \mathcal{B}$. With notation as in Def. 7.2(b), let $b_1 = 1$ and $b_2 = -1$. The condition then implies that $\|x-y\| \leq r+s$ — therefore $B(x,r) \cap B(y,s) \neq \emptyset$ by the observation above. Since $\mathcal{B}$ satisfies this binary intersection property, $\bigcap \mathcal{B} \neq \emptyset$ by Helly's Intersection Theorem 1.2. A similar (but more difficult) argument shows that if $\mathcal{B}$ is a collection of closed balls in $\mathbf{C}$ that satisfies Holbrook's intersection property, then any three of them meet; hence, by Helly's Intersection Theorem 1.2, $\bigcap \mathcal{B} \neq \emptyset$.

Finally, it seems clear to us that the Hahn–Banach theorem should be called, in chronological order, the Helly–Hahn–Banach theorem. Of course, as with so many other misnamed results, this will never happen, a final piece of bad luck for Eduard Helly.

References

1. S. Banach [1929], *Sur les fonctionelles linéaires*, Studia Math. **1**, 211–216 and 223–229, reprinted in Banach [1979] below.
2. S. Banach [1932], *Théorie des opérations linéaires*, Monografje Matematyczne, Warszawa, 1932, reprinted by Chelsea, New York 1932 and in English translation as *Theory of linear operations*, North-Holland, Amsterdam New York Oxford Tokyo, 1987; in this translation all Banach's footnotes and a majority of his references are removed so the history of the subject cannot be properly understood without looking at the Warsaw or New York version. For more on Banach, see http://banach.univ.gda.pl/e-index.html.
3. S. Banach [1979], *Oeuvres*, vol. II, PWN-Éditions Scientifiques de Pologne, Warsaw.
4. S. Banach and S. Mazur [1933], *Zur Theorie der linearen Dimension*, Studia Math. **4**, 100–112, reprinted in Banach [1979] above.
5. H. Bohnenblust and A. Sobczyk [1938], *Extensions of functionals on complex linear spaces*, Bull. Amer. Math. Soc. **44**, 91–93.
6. P. Butzer, S. Gieseler, F. Kaufmann, R. Nessel and E. Stark [1980], *Eduard Helly (1884–1943), Eine nachträgliche Würdigung*, Jahresber. Deutsch. Math.-Verein **82**, 128–151.
7. P. Butzer, R. Nessel and E. Stark [1984], *Eduard Helly (1884–1943), in memoriam*, Results in Mathematics **7**, 145–153.
8. J. Dieudonné [1981], *History of functional analysis*, North-Holland, New York.
9. B. Grünbaum and V. Klee [1967], *Proceedings of the CUPM geometry conference, Part I: Convexity and Applications*, edited by L. Durst, Mathematical Assoc. America, Berkeley, California.

10. H. Hahn [1927], *Über linearer Gleichungssysteme in linearer Räumen*, J. Reine Angew. Math. **157**, 214–229.
11. M. Hasumi [1958], *The extension property of complex Banach spaces*, Tohoku Math. J. **10**, 135–142.
12. E. Helly [1912], *Über linearer Funktionaloperationen*, Österreich. Akad. Wiss. Math.–Natur. Kl. S.-B. IIa, **121**, 265–297.
13. E. Helly [1921], *Über Systeme linearer Gleichungen mit unendlich vielen Unbekannten*, Monatsh. für Math. und Phys., **31**, 60–91.
14. E. Helly [1923], *Über Mengen konvexer Körper mit gemeinschaftlichen Punkten*, Jahresber. Deutsch. Math.–Verein **32**, 175–176.
15. E. Helly [1930], *Über Systeme von abgeschlossenen Mengen mit gemeinschaftlichen Punkten*, Monatsh. für Math. und Phys., **37**, 281–302.
16. W. Helly [1975], *Urban systems models*, Academic Press, New York.
17. J. Holbrook [1975], *Concerning the Hahn–Banach theorem*, Proc. Am. Math. Soc. **50**, 322–327.
18. O. Hustad [1973], *A note on complex $\mathcal{P}_1$-spaces*, Israel J. Math. **16**, 117–119.
19. J. Kelley [1952], *Banach spaces with the extension property*, Trans. Amer. Math. Soc. **72**, 323–326.
20. H. Löwig [1934], *Komplexe euklidische Räume von beliebiger endlicher oder transfiniter Dimensionzahl*, Acta Sci. Math (Szeged) **7**, 1–33.
21. J. Mira [1982], *A unified approach to the extension problem for normed spaces*, Boll. Un. Mat. Ital. **1**, 225–232.
22. F. Murray [1936], *Linear transformations in L_p, $p > 1$*, Trans. Am. Math. Soc. **39**, 83–100.
23. L. Narici [2007], *On the Hahn–Banach theorem*, Advanced Courses of Mathematical Analysis II, Proceedings of the 2nd International School of Analysis in Andalucía 2004, World Scientific Publishing, Singapore 2007, 87-122 or Topology Atlas Preprint #554, http://at.yorku.ca/p/a/a/o/58.htm.
24. L. Nachbin [1950], *A theorem of Hahn–Banach type for linear transformations*, Trans. Am. Math. Soc. **68**, 28–46.
25. J. O'Connor and E. Robertson [2002], *Tibor Radó*, MacTutor History of Mathematics, www-history.mcs.st-andrews.ac.uk/Biographies/Rado.html.
26. T. Ono [1953], *A generalization of the Hahn–Banach theorem*, Nagoya Math. J. **6**, 171–176.
27. J. Radon [1921], *Mengen konvexer Körper, die einen gemeinsamen Punkt enthalten*, Math. Ann. **83**, 113-115.
28. F. Riesz [1910], *Sur certain systèmes d'équations fonctionelles et l'approximation des fonctions continues*, Académie des Sciences, Paris, Comptes Rendus **150**, 674–677 (there were many qualified *émigrés*).
29. F. Riesz [1910a], *Untersuchungen über Systeme integrierbarer Funktionen*, Math. Ann. **69**, 449-497.
30. F. Riesz [1911], *Sur certain systèmes singuliers d'équations intégrales*, Ann. Sci. École Norm. Sup. **28**, 33–62.
31. F. Riesz [1913], *Les systèmes d'équations linéaires à une infinité d'inconnues*, Gauthier-Villars, Paris.

32. F. Riesz [1918], *Über lineare Funktionalgleichungen*, Acta Math. **41**, 71-98. Also in his complete works [1960] below, 1053–1080.
33. F. Riesz [1960], *Oeuvres complètes*, 2 vol., Akadémiai Kiadó, Budapest.
34. E. Schmidt [1908], *Über die Auflösung linearer Gleichungen mit abzählbar unendlich vielen Unbekannten*, Rend. Palermo XXV, 53–77.
35. G. A. Soukhomlinov [1938], *On the extension of linear functionals in complex and quaternion linear spaces*, Matem. Sbornik **3**, 353–358 [Russian with German summary].

THE BANACH SPACE L_p

EDWARD ODELL

Department of Mathematics
The University of Texas at Austin
78712-1082 Austin, TX, USA
E-mail: odell@math.utexas.edu

Our goal is to explore the structure of the "small" subspaces of L_p, mainly for $2 < p < \infty$, discussing older classical results and ultimately presenting some new results of [12]. We will review first some Banach space basics. By L_p we shall mean $L_p[0,1]$, under Lebesgue measure m.

Unless we say otherwise $X, Y, \ldots$ shall denote separable infinite dimensional Banach spaces. $X \subseteq Y$ means that X is a closed subspace of Y. $X \overset{C}{\sim} Y$ means that X is *C-isomorphic to* Y, i.e., there exits an invertible bounded linear $T : X \to Y$ with $\|T\| \, \|T^{-1}\| \leq C$. If $X \overset{1}{\sim} Y$ we shall say X is *isometric* to Y. $X \overset{C}{\hookrightarrow} Y$ means X is C-isomorphic to a subspace of Y.

Definition 1. A *basis* for X is a sequence $(x_i)_1^\infty \subseteq X$ so that for all $x \in X$ there exists a unique sequence $(a_i) \subseteq \mathbb{R}$ with $x = \sum_1^\infty a_i x_i$, i.e., $\lim_n \sum_{i=1}^n a_i x_i = x$.

Example 1. The unit vector basis $(e_i)_{i=1}^\infty$ is a basis for ℓ_p $(1 \leq p < \infty)$. Of course $e_i = (\delta_{i,j})_{j=1}^\infty$ where $\delta_{i,j} = 1$ if $i = j$ and 0 otherwise.

Definition 2. $(x_i)_1^\infty \subseteq X$ is *basic* if $(x_i)_1^\infty$ is a basis for $[(x_i)] \equiv$ the closed linear span of $(x_i)_1^\infty$.

Proposition 1. *Let* $(x_i)_1^\infty \subseteq X$. *Then*

(1) (x_i) *is basic iff* $x_i \neq 0$ *for all* i *and for some* $K < \infty$, *all* $n < m$ *in* $\mathbb{N}$ *and all* $(a_i)_1^m \subseteq \mathbb{R}$,

$$\left\| \sum_1^n a_i x_i \right\| \leq K \left\| \sum_1^m a_i x_i \right\|$$

(In this case (x_i) *is called* K-basic *and the smallest* K *satisfying* (1) *is called the* basis constant *of* (x_i)*).*

(2) (x_i) *is a basis for* X *iff* (1) *holds and* $[(x_i)] = X$.

(x_i) is called *monotone* if its basis constant is 1.

The proof of this and other background facts we present can be found in any of the standard texts such as [3,8,10,25]. The paper [2] contains further background on L_p spaces.

Definition 3. A bounded linear operator $P : X \to X$ is a *projection* if $P^2 = P$.

In this case if $Y = P(X)$ then $X = Y \oplus \text{Ker } P$. Writing $X = Y \oplus Z$ means that Y and Z are closed subspaces of X and every $x \in X$ can be uniquely written $x = y + z$ for some $y \in Y$, $z \in Z$. In this case $Px = y$ defines a projection of X onto Y. Y is said to be *complemented* in X if it is the range of a projection on X. Y is C*-complemented* in X if $\|P\| \le C$.

If $F \subseteq X$ is a finite dimensional subspace then the Hahn–Banach theorem yields that F is complemented in X. If X is isomorphic to ℓ_2 then all $Y \subseteq X$ are complemented but this fails to be the case if $X \not\sim \ell_2$ by a result of Lindenstrauss and Tzafriri ([26]).

Now from Proposition 1 if (x_i) is a basis for X then setting $P_n(\sum a_i x_i) = \sum_{i=1}^n a_i x_i$ yields a projection of X onto $\langle (x_i)_1^n \rangle \equiv$ linear span of $(x_i)_{i=1}^n$. Moreover the P_n's are uniformly bounded and $\sup_n \|P_n\|$ is the basis constant of (x_i).

Not every Banach space X has a basis but the standard ones do.

The Haar basis for L_p $(1 \le p < \infty)$: The Haar basis $(h_i)_1^\infty$ is a monotone basis for L_p.

$$\begin{aligned}
&h_1 = \mathbf{1} \\
&h_2 = \mathbf{1}_{[0,1/2]} - \mathbf{1}_{[1/2,1]} \\
&h_3 = \mathbf{1}_{[0,1/4]} - \mathbf{1}_{[1/4,1/2]} \ , \ h_4 = \mathbf{1}_{[1/2,3/4]} - \mathbf{1}_{[3/4,1]} \\
&\dots
\end{aligned}$$

To see this is a monotone basis for L_p is not hard via Proposition 1. We need only check a couple of things. First note that

$$\left\langle (h_i)_1^{2^n} \right\rangle = \left\{ f = \sum_1^{2^n} a_i \mathbf{1}_{D_i^n} : (a_i)_1^{2^n} \subseteq \mathbb{R} \right\} \quad \text{where} \quad D_i^n = \left[\frac{i-1}{2^n}, \frac{i}{2^n} \right].$$

From real analysis these functions (over all n) are dense in L_p $(1 \le p < \infty)$.

Secondly to see 1) holds with $K = 1$ it suffices to show for all n, $(a_i)_1^{n+1} \subseteq \mathbb{R}$,

$$\left\| \sum_1^n a_i h_i \right\|_p \leq \left\| \sum_1^{n+1} a_i h_i \right\|_p .$$

This reduces to proving if $D = [(i-1)/2^j, i/2^j]$ is a dyadic interval with left half D_+ and right half D_- supporting the Haar function $h = \mathbf{1}_{D_+} - 1_{D_-}$ then for all $a, b \in \mathbb{R}$, $\|a\mathbf{1}_D\|_p \leq \|a\mathbf{1}_D + bh\|_p$. This is easy to see since $a\mathbf{1}_D$ is the average of $a\mathbf{1}_D + bh$ and $a\mathbf{1}_D - bh$, both of which have the same norm.

Definition 4. Basic sequences (x_i) and (y_i) are *C-equivalent* if there exist A, B with $A^{-1}B \leq C$ and for all $(a_i) \subseteq \mathbb{R}$

$$\frac{1}{A} \left\| \sum a_i y_i \right\| \leq \left\| \sum a_i x_i \right\| \leq B \left\| \sum a_i y_i \right\| .$$

This just says that the linear map $T : [(x_i)] \to [(y_i)]$ with $Tx_i = y_i$ for all i is an onto isomorphism with $\|T\| \, \|T^{-1}\| \leq C$.

Proposition 2 (Perturbations). *Let (x_i) be a normalized K-basic sequence in X and let $(y_i) \subseteq X$ satisfy*

$$\sum_{i=1}^{\infty} \|x_i - y_i\| \equiv \lambda < \frac{1}{2K} .$$

Then (x_i) is $C(\lambda)$-equivalent to (y_i) where $C(\lambda) \downarrow 1$ as $\lambda \downarrow 0$. If in addition $[(y_i)]$ is complemented in X by a projection P and $\lambda < 1/(8K\|P\|)$ then $[(x_i)]$ is complemented in X by a projection Q where $\|Q\| \to \|P\|$ as $\lambda \downarrow 0$.

Notation. If (x_i) and (y_i) are C-equivalent basic sequences we write $(x_i) \overset{C}{\sim} (y_i)$.

Definition 5. Let (x_i) be basic. (y_i) is a *block basis* of (x_i) if $y_i \neq 0$ for all i and for some $0 = n_0 < n_1 < n_2 < \cdots$ and $(a_i) \subseteq \mathbb{R}$, $y_i = \sum_{j=n_{i-1}+1}^{n_i} a_j x_j$.

Note. (y_i) is then automatically basic with basis constant not exceeding that of (x_i).

If (x_i) is a normalized K-basis for X we define the *coordinate* or *biorthogonal* functionals (x_i^*) via $x_i^*(\sum a_j x_j) = a_i$. From Proposition 1 we obtain $\|x_i^*\| \leq 2K$ and so for all (a_i)

$$\frac{1}{2K} \|(a_i)\|_\infty \leq \left\| \sum a_i x_i \right\| \leq \sum |a_i| = \|(a_i)\|_{\ell_1} .$$

In other words $\| \sum a_i x_i \|$ is trapped between the c_0 and ℓ_1 norms of (a_i).

From Proposition 2 we obtain

Proposition 3. *Let X have a basis (x_i) and let $(y_i) \subset S_X \equiv \{x \in X : \|x\| = 1\}$ be weakly null (i.e., $x^*(y_i) \to 0$ for all $x^* \in X^*$). Then given $\varepsilon_i \downarrow 0$ there exists a subsequence (z_i) of (y_i) and a block basis $(b_i) \subseteq S_X$ of (x_i) with $\|z_i - b_i\| < \varepsilon_i$ for all i. In particular given $\varepsilon > 0$ we can choose (z_i) to be $(1+\varepsilon)$-equivalent to a normalized block basis of (x_i).*

Definition 6. A basis (x_i) for X is *K-unconditional* if for all $\sum a_i x_i \in X$ and all $\varepsilon_i = \pm 1$,

$$\left\| \sum a_i x_i \right\| \leq K \left\| \sum \varepsilon_i a_i x_i \right\| .$$

It is not hard to show (x_i) is unconditional iff for all $x = \sum a_i x_i \in X$ and all permutations π of $\mathbb{N}$,

$$x = \sum a_{\pi(i)} x_{\pi(i)}$$

iff for some $C < \infty$, all $\sum a_i x_i \in X$ and all $M \subseteq \mathbb{N}$, $\| \sum_{i \in M} a_i x_i \| \leq C \| \sum a_i x_i \|$ (this just says that the projections $(P_M \colon M \subseteq \mathbb{N})$ given by $P_M(\sum a_i x_i) = \sum_{i \in M} a_i x_i$ are well defined and uniformly bounded).

Easily, the unit vector basis (e_i) is a 1-unconditional basis for ℓ_p $(1 \leq p < \infty)$ or c_0.

Example 2. The Haar basis is an unconditional basis for L_p if $1 < p < \infty$.

This is a more difficult result (see [5]), if $p \neq 2$. For $p = 2$, (h_i) is an orthogonal basis

$$\left\| \sum a_i \frac{h_i}{\|h_i\|_2} \right\|_2 = \left(\sum |a_i|^2 \right)^{1/2} .$$

More generally, if (x_i) is a normalized block basis of (h_i) then $\| \sum a_i x_i \|_2 = (\sum |a_i|^2)^{1/2}$.

It is easy to check that (h_i) is not unconditional in L_1. For example if

$$(y_i) = (h_1, h_2, h_3, h_5, h_9, h_{17}, \ldots)$$

is the sequence of "left most" h_i's then

$$\left\| \sum_1^n \frac{y_i}{\|y_i\|_1} \right\|_1 = 1 \text{ while for some } c > 0 , \quad \left\| \sum_1^n (-1)^i \frac{y_i}{\|y_i\|_1} \right\|_1 \geq cn.$$

Definition 7. A *finite dimensional decomposition* (FDD) for X is a sequence of non-zero finite dimensional subspaces (F_i) of X so that for all $x \in X$ there exists a unique sequence (x_i) with $x_i \in F_i$ for all i and $x = \sum x_i$.

As with bases, the projections $P_n x = P_n(\sum x_i) = \sum_1^n x_i$ are uniformly bounded and $\sup_n \|P_n\|$ is the basis constant of the FDD. Also for $n \le m$ if $P_{[n,m]}x = \sum_n^m x_i$, then the $P_{[n,m]}$'s are uniformly bounded and $\sup_{n\le m} \|P_{[n,m]}\|$ is the *projection constant* of the FDD. (E_i) is monotone if its basis constant is 1 and *bimonotone* if its projection constant is 1.

A *blocking* (G_i) of an FDD (F_i) for X is given by $G_i = \langle (F_j)_{j=n_{i-1}+1}^{n_i} \rangle$ for some $0 = n_0 < n_1 < \cdots$. (G_i) is then also an FDD for X.

A basis (x_i) also may be regarded as an FDD with $F_i = \langle x_i \rangle$.

From Proposition 3 we see that if $1 < p < \infty$ and $(y_i) \subseteq S_{L_p}$ is weakly null (equivalently, $\int_E y_i \to 0$ for all measurable $E \subseteq [0,1]$) then some subsequence is a perturbation of a block basis of (h_i) and hence is unconditional (just like for bases, block bases of unconditional bases are unconditional). This fails in L_1 by a deep new result of Johnson, Maurey and Schechtman.

Theorem 1 ([14]). *There exists a weakly null sequence $(x_i) \subseteq S_{L_1}$ satisfying: for all $\varepsilon > 0$ and all subsequences $(y_i) \subseteq (x_i)$, (h_i) is $(1+\varepsilon)$-equivalent to a block basis of (y_i).*

Now let's fix $2 < p < \infty$ and let K_p be the unconditional constant of (h_i) in L_p. We shall list what we consider to be the small subspaces of L_p. These are also subspaces of L_p for $1 < p < 2$ but as we shall note shortly the situation there as to what constitutes "small" is more complicated.

L_p contains the following "small" subspaces:

- ℓ_p (isometrically): If $(x_i) \subseteq S_{L_p}$ are disjointly supported, $(|x_i| \wedge |x_j| = 0$ for $i \ne j)$ then

$$\begin{aligned}\left\|\sum a_i x_i\right\| &= \left(\int |\sum a_i x_i(t)|^p\, dt\right)^{1/p} \\ &= \left(\sum \int |a_i|^p |x_i(t)|^p\, dt\right)^{1/p} \\ &= \left(\sum |a_i|^p\right)^{1/p}.\end{aligned}$$

Also $[(x_i)]$ is 1-complemented in X via $Px = \sum_{i=1}^\infty x_i^*(x) x_i$ where (x_i^*) are the functions naturally biorthogonal to (x_i), $x_i^* = \operatorname{sign}(x_i)|x_i|^{p-1}$.

- ℓ_2 (isomorphically) via the Rademacher functions (r_n). (r_n) are ± 1 valued independent random variables of mean 0.

Khintchin's inequality: For $2 < p < \infty$,

$$\left(\sum |a_n|^2\right)^{1/2} = \left\|\sum a_n r_n\right\|_2 \le \left\|\sum a_n r_n\right\|_p \le B_p \left(\sum |a_n|^2\right)^{1/2}.$$

For $1 < p < 2$

$$A_p \left(\sum |a_n|^2\right)^{1/2} \le \left\|\sum a_n r_n\right\|_p \le \left\|\sum a_n r_n\right\|_2 = \left(\sum |a_n|^2\right)^{1/2}.$$

The constants A_p, B_p depend solely on p.

- ℓ_2 (isometrically) via a sequence of symmetric Gaussian independent random variables in S_{L_p}.
- $(\ell_2 \oplus \ell_p)_p$ (isometrically).

For this we use that $L_p \overset{1}{\sim} (L_p[0,1/2] \oplus L_p[1/2,1])_p$ and $L_p[0,1/2] \overset{1}{\sim} L_p[1/2,1] \overset{1}{\sim} L_p[0,1]$. More generally, if we partition $[0,1]$ into disjoint intervals of positive measure $(I_n)_{n=1}^\infty$ then $L_p(I_n) \overset{1}{\sim} L_p$ and $L_p \overset{1}{\sim} (\sum L_p(I_n))_p$. Hence L_p contains also

- $(\sum \ell_2)_p = (\ell_2 \oplus \ell_2 \oplus \cdots)_p \equiv \{(x_i) : x_i \in \ell_2 \text{ for all } i \text{ and } \|(x_i)\| = (\sum \|x_i\|_2^p)^{1/p} < \infty\}$ (isometrically)

Our topic will be to characterize when $X \subseteq L_p$, $2 < p < \infty$, embeds isomorphically into or contains isomorphically one of the four spaces ℓ_p, ℓ_2, $\ell_p \oplus \ell_2$ or $(\sum \ell_2)_p$.

Now some remarks are in order here. First it is known that $L_q \overset{1}{\hookrightarrow} L_p$ if $p < q \le 2$. Thus L_p contain ℓ_q if $p < q < 2$ so is this "small"? Is L_q small? Secondly we have

Proposition 4. *Let $X \subseteq \ell_p$ $(1 \le p < \infty)$. Then for all $\varepsilon > 0$ there exists $Y \subseteq X$ with $Y \overset{1+\varepsilon}{\sim} \ell_p$ and Y is $1+\varepsilon$-complemented in ℓ_p.*

This is due to Pełczyński ([31]). Every normalized block basis of (e_i) in ℓ_p is 1-equivalent to (e_i) and 1-complemented in ℓ_p by the analogous statement in L_p realizing (e_i) as a disjointly supported sequence in S_{L_p}. Then one uses perturbation as in Proposition 3.

Some other classical facts are:

(i) The ℓ_p spaces are totally incomparable, i.e., for all $X \subseteq \ell_p$, $Y \subseteq \ell_q$ if $p \ne q$ then $X \not\sim Y$.

(ii) For $1 \le p, q < \infty$, $L_q \hookrightarrow L_p$ iff $q = 2$ or $1 \le p \le q < 2$. Also $\ell_q \hookrightarrow L_p$ iff $1 \le p \le q < 2$ or $q = 2$.

Our next result shows that normalized unconditional basic sequences in L_p, $1 < p < \infty$, are trapped between the ℓ_p and ℓ_2 norms.

Proposition 5.

(a) Let $2 < p < \infty$ and let $(x_i) \subseteq S_{L_p}$ be λ-unconditional. Then for all $(a_n) \subseteq \mathbb{R}$,

$$\lambda^{-1}\left(\sum |a_n|^p\right)^{1/p} \le \left\|\sum a_n x_n\right\|_p \le \lambda B_p \left(\sum |a_n|^2\right)^{1/2}.$$

(b) Let $1 < p < 2$ and let $(x_i) \subseteq S_{L_p}$ be λ-unconditional. Then for all $(a_i) \subseteq \mathbb{R}$,

$$(\lambda A_p)^{-1}\left(\sum |a_n|^2\right)^{1/2} \le \left\|\sum a_n x_n\right\|_p \le \lambda \left(\sum |a_n|^p\right)^{1/p}.$$

Proof. For $t \in [0,1]$, $2 < p < \infty$,

$$\left\|\sum a_n x_n\right\|_p \le \lambda \left\|\sum a_n x_n r_n(t)\right\|_p$$

and so

$$\begin{aligned}
\left\|\sum a_n x_n\right\|_p^p &\le \lambda^p \int_0^1 \left\|\sum a_n x_n r_n(t)\right\|_p^p dt \\
&\overset{\text{(Fubini)}}{=} \lambda^p \int_0^1 \int_0^1 \left|\sum a_n x_n(s) r_n(t)\right|^p dt\, ds \\
&\le (\lambda B_p)^p \int_0^1 \left(\sum a_n^2 x_n(s)^2\right)^{p/2} ds \\
&\le (\lambda B_p)^p \left(\sum \|a_n^2 x_n^2\|_{p/2}\right)^{p/2} \\
&\overset{\text{(triangle inequality)}}{=} (\lambda B_p)^p \left(\sum |a_n|^2\right)^{p/2}.
\end{aligned}$$

This gives the upper ℓ_2-estimate.

Similarly,

$$\begin{aligned}
\lambda^p \left\|\sum a_n x_n\right\|^p &\ge \int_0^1 \left(\sum a_n^2 x_n^2(s)\right)^{p/2} ds \\
&\ge \int_0^1 \sum |a_n|^p |x_n(s)|^p\, ds = \sum |a_n|^p
\end{aligned}$$

(using $\|\cdot\|_{\ell_p} \le \|\cdot\|_{\ell_2}$). The argument is similar for $1 < p < 2$. $\square$

The technique of proof, integrating against the Rademacher functions, yields

Proposition 6. *For $1 < p < \infty$ there exists $C(p)$ so that if $(x_i) \subseteq S_{L_p}$ is λ-unconditional then for all (a_i)*

$$\Big\|\sum a_n x_n\Big\|_p \overset{\lambda C(p)}{\sim} \left(\int_0^1 \left(\sum |a_n|^2 |x_n(s)|^2\right)^{p/2} ds\right)^{1/p} . \qquad (1)$$

The expression on the right is the so called "square function". By $A \overset{C}{\sim} B$ we mean $A \le CB$ and $B \le CA$. Proposition 6 leads to a very nice result (which we won't need below but is too pretty not to give).

Corollary 1 ([35]). *Let $(x_n) \subseteq S_{L_p}$, $1 < p < \infty$, be unconditional basic. Then (x_n) is equivalent to a block basis (y_n) of (h_n).*

Let's give a brief sketch of the proof. By (1) it follows that if (y_i) is a block basis of (h_i) with $|y_i| = |x_i|$ on $[0,1]$ then $(y_i) \sim (x_i)$. By a perturbation argument we may assume each $x_i \in \langle h_j \rangle$. Then it is easy to construct the y_i's. Indeed given a simple dyadic function x and any n one can find $y \in \langle h_i \rangle_n^\infty$ so that $|y| = |x|$.

We are now ready to begin our investigation announced previously: if $X \subseteq L_p$ $(2 < p < \infty)$, when does X contain or embed into one of the 4 small subspaces of L_p, namely ℓ_p, ℓ_2, $\ell_p \oplus \ell_2$ or $(\sum \ell_2)_p$? We begin with a result from 1960.

Theorem 2 (Kadets and Pełczyński [20]). *Let $X \subseteq L_p$, $2 < p < \infty$. Then $X \sim \ell_2$ iff $\|\cdot\|_2 \sim \|\cdot\|_p$ on X; i.e., for some C, $\|x\|_2 \le \|x\|_p \le C\|x\|_2$ for all $x \in X$. Moreover there is a projection $P\colon L_p \to X$.*

First note that if $x \in S_{L_p}$ and $m\{t : |x(t)| \ge \varepsilon\} \ge \varepsilon$ then $\|x\|_2 \le \|x\|_p = 1 \le \varepsilon^{-3/2}\|x\|_2$. Indeed

$$\|x\|_2 = \left(\int |x(t)|^2\, dt\right)^{1/2} \ge \left(\int_{[|x| \ge \varepsilon]} |x(t)|^2\, dt\right)^{1/2} \ge \varepsilon \cdot \varepsilon^{1/2} .$$

The direction requiring proof is if $X \sim \ell_2$ then $\|\cdot\|_2 \sim \|\cdot\|_p$ on X. If not we can find $(x_i) \subseteq S_X$, $x_i \xrightarrow{\omega} 0$, so that for all $\varepsilon > 0$, $\lim_n m[|x_n| \ge \varepsilon] = 0$. From this we can construct a subsequence (x_{n_i}) and disjointly supported $(f_i) \subseteq S_{L_p}$ with $\lim_i \|x_{n_i} - f_i\| = 0$. Hence by a perturbation argument a subsequence of (x_i) is equivalent to the unit vector basis of ℓ_p which contradicts $X \sim \ell_2$. The projection onto X with $\|x\|_p \le C\|x\|_2$ for $x \in X$ is given by the orthogonal projection $P\colon L_2 \to X$ acting on L_p. For $y \in L_p$,

$$\|Py\|_p \le C\|Py\|_2 \le C\|y\|_2 \le C\|y\|_p.$$

Remark 1. The proof yields that if $X \subseteq L_p$, $2 < p < \infty$, and $X \not\sim \ell_2$ then for all $\varepsilon > 0$, $\ell_p \overset{1+\varepsilon}{\hookrightarrow} X$. Moreover if $(x_i) \subseteq S_{L_p}$ is weakly null and $\varepsilon = \lim_i \|x_i\|_2$ then a subsequence is equivalent to the ℓ_p basis if $\varepsilon = 0$ and the ℓ_2 basis if $\varepsilon > 0$.

In the latter case we have essentially by Proposition 5 (assuming say (x_i) is a normalized block basis of (h_i) with $\|x_i\|_2 = \varepsilon$ for all i)

$$\varepsilon\left(\sum a_i^2\right)^{1/2} = \left\|\sum a_i x_i\right\|_2 \le \left\|\sum a_i x_i\right\|_p \le K_p B_p \left(\sum a_i^2\right)^{1/2}.$$

Pełczyński and Rosenthal ([32]) proved that if $X \overset{K}{\sim} \ell_2$ then X is $C(K)$-complemented in L_p via a change of density argument.

Our next result shows that if X does not contain an isomorph of ℓ_2 then it embeds into ℓ_p. The argument uses "Pełczyński's decomposition method".

Proposition 7 ([31]). *Let X be a complemented subspace of ℓ_p, $1 \le p < \infty$. Then $X \sim \ell_p$.*

Proof. $\ell_p \sim X \oplus V$ for some $V \subseteq \ell_p$. Also $X \sim \ell_p \oplus W$ for some $W \subseteq X$ by Proposition 4. Finally $\ell_p \sim \ell_p \oplus \ell_p$ and moreover $\ell_p \sim (\ell_p \oplus \ell_p \oplus \cdots)_p$. The latter is proved by splitting (e_i) into infinitely many infinite subsets. Thus

$$\begin{aligned}
\ell_p &\sim (\ell_p \oplus \ell_p \oplus \cdots)_p \sim ((X \oplus V) \oplus (X \oplus V) \oplus \cdots)_p \\
&\sim (X \oplus X \oplus \cdots)_p \oplus (V \oplus V \oplus \cdots)_p \\
&\sim X \oplus (X \oplus X \oplus \cdots)_p \oplus (V \oplus V \oplus \cdots)_p \\
&\sim X \oplus \ell_p \sim W \oplus \ell_p \oplus \ell_p \sim W \oplus \ell_p \sim X\,.
\end{aligned}$$

□

A consequence of this is that if (H_n) is any blocking of (h_i) into an FDD then $(\sum H_n)_p \sim \ell_p$. Indeed each H_n is uniformly complemented in L_p and hence in $\ell_p^{m_n}$ for some m_n. Thus $(\sum H_n)_p$ is complemented in $(\sum \ell_p^{m_n})_p = \ell_p$.

Theorem 3 ([16]). *Let $2 < p < \infty$, $X \subseteq L_p$. Then $X \hookrightarrow \ell_p \Leftrightarrow \ell_2 \not\hookrightarrow X$. ([21] If $\ell_2 \not\hookrightarrow X$ then for all $\varepsilon > 0$, $X \overset{1+\varepsilon}{\hookrightarrow} \ell_p$.)*

The scheme of the argument is to show if $\ell_2 \not\hookrightarrow X$ then there is a blocking (H_n) of the Haar basis into an FDD so that $X \hookrightarrow (\sum H_n)_p$ in a natural way; $x = \sum x_n$, $x_n \in H_n \to (x_n) \in (\sum H_n)_p$. Since $(\sum H_n)_p \sim \ell_p$ we are done.

We won't discuss the specifics here of this argument but rather will sketch shortly the proof of a stronger result. First we note the analogous theorem for $1 < p < 2$, which necessarily has a different form. Note Theorem 4 would also hold for $2 < p < \infty$ and, unlike $1 < p < 2$, the constant K need not be specified.

Theorem 4 ([13]). *Let $X \subseteq L_p$, $1 < p < 2$. Then $X \hookrightarrow \ell_p$ if (and only if) there exists $K < \infty$ so that for all weakly null $(x_i) \subseteq S_X$ some subsequence is K-equivalent to the unit vector basis of ℓ_p.*

These results were unified using the infinite asymptotic game/weakly null trees machinery which we will discuss after stating

Theorem 5. *Let $X \subseteq L_p$, $1 < p < \infty$. Then $X \hookrightarrow \ell_p$ iff every weakly null tree in S_X admits a branch equivalent to the unit vector basis of ℓ_p.*

A tree in S_X is $(x_\alpha)_{\alpha \in T_\infty} \subseteq S_X$ where

$$T_\infty = \{(n_1, \ldots, n_k) : k \in \mathbb{N},\ n_1 < \cdots < n_k \text{ are in } \mathbb{N}\}.$$

A *node* in T_∞ is all $(x_{(\alpha,n)})_{n>n_k}$ where $\alpha = (n_1, \ldots, n_k)$ or $\alpha = \emptyset$. The tree is *weakly null* means each node is a weakly null sequence. A *branch* is $(x_i)_{i=1}^\infty$ given by $x_i = x_{(n_1,\ldots,n_i)}$ for some subsequence (n_i) of $\mathbb{N}$.

It is worth noting that, just as in Proposition 3, if $X \subseteq Z$, a space with a basis (z_i), and $(x_\alpha)_{\alpha \in T_\infty} \subseteq S_X$ is a weakly null tree then the tree admits a full subtree $(y_\alpha)_{\alpha \in T_\infty}$ so that each branch is a perturbation of a block basis of (z_i). By *full subtree* we mean that $(y_\alpha)_{\alpha \in T_\infty} = (x_\alpha)_{\alpha \in T'}$ where $T' \subseteq T_\infty$ is order isomorphic to T_∞ and if $y_\alpha = x_{\gamma(\alpha)}$ then $|\gamma(\alpha)| = |\alpha| =$ length of α. $|(n_1, \ldots, n_k)| = k$. Thus each branch of $(y_\alpha)_{\alpha \in T_\infty}$ is a branch of $(x_\alpha)_{\alpha \in T_\infty}$.

Remark 2. The conditions for a reflexive space X,

A) Every weakly null sequence $(x_i) \subseteq X$ has a subsequence K-equivalent to the unit vector basis of ℓ_p and

B) Every weakly null tree in S_X admits a branch equivalent to the unit vector basis of ℓ_p are generally different. Also it is not hard to show that B) actually implies

B)$'$ For some C every weakly null tree in S_X admits a branch C-equivalent to the unit vector basis of ℓ_p.

Note that B)$'$ $\Rightarrow$A) by considering for a weakly null $(x_n) \subseteq S_X$, the tree $(x_\alpha)_{\alpha \in T_\alpha}$ where $x_{(n_1,\dots,n_k)} = x_{n_k}$. Indeed the branches of (x_α) coincide with the subsequences of (x_i). But in L_p one can show that A) and B) are in fact equivalent. Thus Theorem 5 encompasses both Theorems 3 and 4.

Theorem 5 follows from

Theorem 6 ([30]). *Let $1 < p < \infty$, let X be reflexive and assume that every weakly null tree in S_X admits a branch C-equivalent to the unit vector basis of ℓ_p. Assume $X \subseteq Z$, a reflexive space with an FDD(E_i). Then there exists a blocking (F_i) of (E_i) so that X naturally embeds into $(\sum F_i)_p$.*

The conclusion means that for some K and all $x \in X$, $x = \sum x_n$, $x_n \in F_n$, we have $\|x\| \overset{K}{\sim} (\sum \|x_n\|^p)^{1/p}$.

We shall outline the steps involved in the proof. First we give a definition.

Definition 8. Let (E_i) be an FDD for Z. Let $\bar{\delta} = (\delta_i)$, $\delta_i \downarrow 0$. A sequence $(z_i) \subseteq S_Z$ is a *$\bar{\delta}$-skipped block sequence* w.r.t. (E_i) if there exist integers $1 \le k_1 < \ell_1 < k_2 < \ell_2 < \cdots$ so that

$$\|z_n - P^E_{(k_n,\ell_n]} z_n\| < \delta_n \quad \text{for all} \quad n.$$

Here for $x = \sum x_i$, $x_i \in E_i$, $P^E_{(k,\ell]}x = \sum_{i \in (k,\ell]} x_i$. Thus above the "skipping" is the $P^E_{k_n}$ terms. (z_n) is almost a block basis of (E_n) with the E_{k_n} almost skipped.

Now let $X \subseteq Z = [(E_i)]$ be as in the statement of Theorem 6.

Step 1. There exists a blocking (G_i) of (E_i) and $\bar{\delta}$ so that every $\bar{\delta}$-skipped block sequence w.r.t. (G_i) in S_X is $2C$-equivalent to the unit vector basis of ℓ_p.

To obtain this one first shows that the weakly null tree hypothesis on X is equivalent to (S) having a winning strategy in the following game (for all $\varepsilon > 0$).

> *The infinite asymptotic game*: Two players (S) for subspace and (V) for vector alternate plays forever as follows. (S) chooses $n_1 \in \mathbb{N}$. (V) chooses $x_1 \in S_X \cap [(E_i)_{i \ge n_1}], \dots$. Thus the plays are $(n_1, x_1, n_2, x_2, \dots)$.

(S) *wins* if $(x_i) \in \mathcal{A}(C + \varepsilon) \equiv \{$all normalized bases $(C + \varepsilon)$-equivalent to the unit vector basis of $\ell_p\}$.

(S) *has a winning strategy* means that

$$\begin{aligned}
&\exists\, n_1\ \forall\, x_1 \in S_X \cap [(E_i)_{i\ge n_1}]\\
&\exists\, n_2\ \forall\, x_2 \in S_X \cap [(E_i)_{i\ge n_2}]\\
&\cdots\\
&(x_i) \in \mathcal{A}(C+\varepsilon)
\end{aligned}$$

(V) *wins* if $(x_i) \notin \mathcal{A}(C+\varepsilon)$.

(V) *has a winning strategy* means that

$$\begin{aligned}
&\forall\, n_1\ \exists\, x_1 \in S_X \cap [(E_i)_{i\ge n_1}]\\
&\forall\, n_2\ \exists\, x_2 \in S_X \cap [(E_i)_{i\ge n_2}]\\
&\cdots\\
&(x_i) \notin \mathcal{A}(C+\varepsilon)
\end{aligned}$$

Now these two winning strategies are the formal negations of each other, but they are infinite sentences so must one be true? Yes, if the game is determined which it is in this case since Borel games are determined ([27]). Now if (V) had a winning strategy one could easily produce a weakly null tree in S_X all of whose branches did not lie in $\mathcal{A}(C+\varepsilon)$. So (S) has a winning strategy. Then by a compactness argument one can deduce Step 1 ($2C$ could be any $C+\varepsilon$ here).

The next step is a lemma of W.B. Johnson ([13]) which allows us to decompose any $x \in S_X$ into (almost) a linear combination of $\bar{\delta}$-skipped blocks, in X.

Step 2. Let K be the projection constant of (G_i). There exists a blocking (F_i) of (G_i), $F_i = \langle G_i\rangle_{j\in(N_{i-1},N_i]}$, $N_0 = 0 < N_1 < \cdots$, satisfying the following.

For all $x \in S_X$ there exists $(x_i) \subseteq X$ and for all i there exists $t_i \in (N_{i-1}, N_i)$ ($t_0 = 0$, $t_1 > 1$) satisfying

(a) $x = \sum x_j$
(b) $\|x_i\| < \delta_i$ or $\|P^G_{(t_{i-1},t_i)} x_i - x_i\| < \delta_i \|x_i\|$
(c) $\|P^G_{(t_{i-1},t_i)} x - x_i\| < \delta_i$
(d) $\|x_i\| < K+1$
(e) $\|P^G_{t_i} x\| < \delta_i$

Moreover the above holds for any further blocking of (G_i) (which redefines the N_i's).

Remark 3. Thus if $x \in S_X$ we can write $x = \sum x_i$, $(x_i) \subseteq X$ where if $B = \{i \geq 2 : \|x_i\| \geq \delta_i\}$ then $(x_i/\|x_i\|)_{i\in B}$ is a $\bar{\delta}$-skipped block sequence w.r.t. (G_i). Also the skipped blocks (G_{t_i}) are in predictable intervals, $t_i \in (N_{i-1}, N_i)$. And $\sum_{\substack{i\notin B\\ i>1}} \|x_i\| < \sum \delta_i$.

To prove Step 2 we have a

Lemma 1. *$\forall\ \varepsilon > 0\ \forall\ N \in \mathbb{N}\ \exists\ n > N$ so that if $x \in B_X$, $x = \sum y_i$, $y_i \in G_i$, then there exists $t \in (N, n)$ with*

$$\|y_t\| < \varepsilon \quad \textit{and} \quad \operatorname{dist}\left(\sum_1^{t-1} y_i, X\right) < \varepsilon.$$

Proof. If not we obtain $y^{(n)} \in B_X$ for $n > N$ failing the conclusion for $t \in (N, n)$. Choose $y^{(n_i)} \xrightarrow{\omega} y \in B_X$ and let $t > N$ satisfy $\|P^G_{[t,\infty)} y\| < \varepsilon/2K$. Choose $y^{(n)}$ from $(y^{(n_i)})$ so that $n > t$ and $\|P^G_{[1,t)}(y^{(n)} - y)\| < \varepsilon/2K$. Then

$$\|P^G_{[1,t)} y^{(n)} - y\| \leq \|P^G_{[1,t)}(y^{(n)} - y)\| + \|P^G_{[t,\infty)} y\| < \frac{\varepsilon}{2K} + \frac{\varepsilon}{2K} \leq \varepsilon.$$

Also

$$\|P^G_t y^{(n)}\| \leq \|P^G_t (y^{(n)} - y)\| + \|P^G_t y\| < \frac{\varepsilon}{2} + \frac{\varepsilon}{2} = \varepsilon.$$

This contradicts our choice of $y^{(n)}$. □

To use the lemma we select $N_0 = 0 < N_1 < N_2 < \cdots$ so that for all $x \in B_X$ there exists $t_i \in (N_{i-1}, N_i)$ and $z_i \in X$ with $\|P^G_{t_i} x\| < \varepsilon_i$ and $\|P^G_{[1,t_i)} x - z_i\| < \varepsilon_i$. Set $x_i = z_1$, $x_i = z_i - z_{i-1}$ for $i > 1$. Then $\sum_1^n x_i = z_n \to x$ and the other properties b)–d) hold, as is easily checked, if $(K+1)(\varepsilon_i + 2\varepsilon_{i-1}) < \delta_i^2$.

Now let (F_i) be the blocking obtained in Step 2. It is not hard to show that if $x = \sum x_i$ is as in Step 2 then $\|x\| \overset{3C}{\sim} (\sum \|x_i\|^p)^{1/p}$, provided $\bar{\delta} = (\delta_i)$ is small enough. But this is not the decomposition given by $x = \sum y_i$, $y_i \in F_i$. However we do have

$$\begin{aligned} x_i &\approx P^G_{(t_{i-1}, t_i)}(y_{i-1} + y_i) \quad \text{and} \\ y_i &\approx P^G_{(N_{i-1}, N_i)}(x_i + x_{i+1}) \end{aligned}$$

which yields $\|x\| \overset{K(C)}{\sim} (\sum \|y_i\|^p)^{1/p}$ by making the appropriate estimates.

Returning to $X \subseteq L_p$ $(2 < p < \infty)$ we have seen that one of these holds:

- $X \sim \ell_2$

- $X \hookrightarrow \ell_p$
- $\ell_p \oplus \ell_2 \hookrightarrow X$

The latter comes from Theorems 2 and 3. If $X \not\sim \ell_2$ and $X \not\hookrightarrow \ell_p$ then X contains a subspace isomorphic to ℓ_2 so $X \sim \ell_2 \oplus Y$. Now Y also contains ℓ_p (or else $X \sim \ell_2$) and in fact complementably (as a perturbation of a disjointly supported $(f_i) \subseteq S_{L_p}$) so $\ell_p \oplus \ell_2 \hookrightarrow X$.

Our next goal will be to characterize when $X \hookrightarrow \ell_p \oplus \ell_2$ and if not to then show that $(\sum \ell_2)_p \hookrightarrow X$.

First we recall one more old result.

Theorem 7 ([17]). *Let $X \subseteq L_p$, $2 < p < \infty$. Assume there exists $Y \subseteq \ell_p \oplus \ell_2$ and a quotient (onto) map $Q\colon Y \to X$. Then $X \hookrightarrow \ell_p \oplus \ell_2$.*

This is an answer, of a sort, to when $X \hookrightarrow \ell_p \oplus \ell_2$ but it is not an intrinsic characterization. The proof however provides a clue as to how to find one. The isomorphism $X \hookrightarrow \ell_p \oplus \ell_2$ is given by a blocking (H_n) of (h_i) so that X naturally embeds into

$$\Big(\sum H_n\Big)_p \oplus \Big(\sum (H_n, \|\cdot\|_2)\Big)_2 \sim \ell_p \oplus \ell_2.$$

Before proceeding we recall some more inequalities.

Theorem 8 ([33]). *Let $2 < p < \infty$. There exists $K_p < \infty$ so that if (x_i) is a normalized mean zero sequence of independent random variables in L_p then for all $(a_i) \subseteq \mathbb{R}$,*

$$\Big\|\sum a_i x_i\Big\|_p \overset{K_p}{\sim} \Big(\sum |a_i|^p\Big)^{1/p} \vee \Big(\sum |a_i|^2 \|x_i\|_2^2\Big)^{1/2}.$$

Note that in this case $[(x_i)] \hookrightarrow \ell_p \oplus \ell_2$ via the embedding

$$\sum a_i x_i \longmapsto ((a_i)_i, (a_i \|x_i\|_2)_i) \in \ell_p \oplus \ell_2.$$

The next result generalizes this to martingale difference sequences, e.g., block bases of (h_i).

Theorem 9 ([6,7]). *Let $2 < p < \infty$. There exists $C_p < \infty$ so that if (z_i) is a martingale difference sequence in L_p with respect to the sequence of σ-algebras $(\mathcal{F}_n)$, then*

$$\Big\|\sum z_i\Big\|_p \overset{C_p}{\sim} \Big(\sum \|z_i\|_p^p\Big)^{1/p} \vee \Big\|\Big(\sum \mathbb{E}_{\mathcal{F}_i}(z_{i+1}^2)\Big)^{1/2}\Big\|_p.$$

Recall something we said earlier. Suppose that $(x_i) \subseteq S_{L_p}$ is weakly null. Passing to a subsequence we obtain (y_i) which, by perturbing, we may assume is a block basis of (h_i). Passing to a further subsequence we may assume $\varepsilon \equiv \lim_i \|y_i\|_2$ exists. If $\varepsilon = 0$ a subsequence of (y_i) is equivalent to the unit vector basis of ℓ_p by the [20] arguments. Otherwise we have (essentially)

$$\varepsilon \left(\sum |a_i|^2\right)^{1/2} = \left\|\sum a_i y_i\right\|_2 \leq \left\|\sum a_i y_i\right\|_p \leq C(p) \left(\sum |a_i|^2\right)^{1/2},$$

using the fundamental inequality, Proposition 5. Thus $[(y_i)]$ embeds into $\ell_p \oplus \ell_2$ with (y_i) as a block basis of the natural basis for $\ell_p \oplus \ell_2$.

Johnson, Maurey, Schechtman and Tzafriri obtained a stronger version of this dichotomy using Theorem 9.

Theorem 10 ([15]). *Let $2 < p < \infty$. There exists $D_p < \infty$ with the following property. Every normalized weakly null sequence in L_p admits a subsequence (x_i) satisfying, for some $w \in [0,1]$ and all $(a_i) \subseteq \mathbb{R}$,*

$$\left\|\sum a_i x_i\right\|_p \overset{D_p}{\sim} \left(\sum |a_i|^p\right)^{1/p} \vee w \left(\sum |a_i|^2\right)^{1/2}.$$

We are now ready for an intrinsic characterization of when $X \subseteq L_p$ embeds into $\ell_p \oplus \ell_2$.

Theorem 11 ([12]). *Let $X \subseteq L_p$, $2 < p < \infty$. The following are equivalent:*

(a) $X \hookrightarrow \ell_p \oplus \ell_2$
(b) Every weakly null tree in S_X admits a branch (x_i) satisfying for some K and all (a_i)

$$\left\|\sum a_i x_i\right\| \overset{K}{\sim} \left(\sum |a_i|^p\right)^{1/p} \vee \left\|\sum a_i x_i\right\|_2$$

$$\approx \left(\sum |a_i|^p\right)^{1/p} \vee \left(\sum |a_i|^2 \|x_i\|_2^2\right)^{1/2}.$$

(c) Every weakly null tree in S_X admits a branch (x_i) satisfying for some K and $(w_i) \subseteq [0,1]$ and all (a_i),

$$\left\|\sum a_i x_i\right\| \overset{K}{\sim} \left(\sum |a_i|^p\right)^{1/p} \vee \left(\sum |a_i|^2 w_i^2\right)^{1/2}.$$

(d) *There exists K so that every weakly null sequence in S_X admits a subsequence (x_i) satisfying the condition in b):*

$$\left\|\sum a_i x_i\right\| \overset{K}{\sim} \left(\sum |a_i|^p\right)^{1/p} \vee \left\|\sum a_i x_i\right\|_2$$

$$\approx \left(\sum |a_i|^p\right)^{1/p} \vee \left(\sum |a_i|^2 \varepsilon^2\right)^{1/2}$$

$$\textit{where} \quad \varepsilon = \lim_i \|x_i\|_2 .$$

Condition c) just says that every weakly null tree in S_X admits a branch equivalent to a block basis of the natural basis for $\ell_p \oplus \ell_2$ (discussed more below).

Conditions b) and c) do not require K to be universal but the "all weakly null trees..." hypothesis yields this.

The latter "$\approx$" near equalities in b) (and d)) come from the fact that every weakly null tree in S_{L_p} can be first pruned to a full subtree so that each branch is essentially a normalized block basis of (h_i).

Condition d) is an anomaly in that usually "every sequence has a subsequence..." is a vastly different condition than "every tree admits a branch...". Here again the special nature of L_p is playing a role. Also note the difference between d) and Theorem 10.

The embedding of X into $\ell_p \oplus \ell_2$ will follow the clue from the proof of Theorem 7 by producing a blocking (H_n) of (h_i) and embedding X naturally into

$$\left(\sum H_n\right)_p \oplus \left(\sum (H_n, \|\cdot\|_2)\right)_2 .$$

Thus if $x = \sum x_n$, $x_n \in H_n$ then $\|x\| \sim (\sum \|x_n\|^p)^{1/p} \vee (\sum \|x_n\|_2^2)^{1/2}$.

The proof of b)$\Rightarrow$a) is much like that of Theorem 6. We produce a blocking (H_n) of (h_n) so that X naturally embeds into $(\sum H_n)_p \oplus (\sum (H_n, \| \cdot \|_2))_2 \sim \ell_p \oplus \ell_2$. In fact we obtain a more general result.

A basis (v_i) is 1-*subsymmetric* if it is 1-unconditional and $\|\sum a_i v_i\| = \|\sum a_i v_{n_i}\|$ for all (a_i) and all $n_1 < n_2 < \cdots$.

Theorem 12 ([12]). *Let X and Y be Banach spaces with X reflexive. Let V be a space with a 1-subsymmetric normalized basis (v_i) and let $T\colon X \to Y$ be a bounded linear operator. Assume that for some C every normalized weakly null tree in X admits a branch (x_n) satisfying:*

$$\left\|\sum a_n x_n\right\|_X \overset{C}{\sim} \left\|\sum a_n v_n\right\|_V \vee \left\|T\left(\sum a_n x_n\right)\right\|_Y .$$

Then if $X \subseteq Z$, a reflexive space with an FDD(E_i), *there exists a blocking (G_i) of (E_i) so that X naturally embeds into $(\sum G_i)_V \oplus Y$: if $x = \sum x_i$, $x_i \in G_i$ then $x \mapsto (x_i) \oplus Tx \in (\sum G_i)_V \oplus Y$.*

$(\sum G_i)_V$ is the completion of $\langle (G_i) \rangle$ under

$$\|(x_i)\| = \left\| \sum \|x_i\| v_i \right\|_V .$$

This is applied to $V = \ell_p$, $Z = L_p$ and $Y = L_2$ where $T : X \to L_2$ is the identity map.

So we obtain b)⇒a) and clearly a)⇒c). Indeed suppose that $X \subseteq (\ell_p \oplus \ell_2)_\infty$. Then given a weakly null tree in X some branch (x_i) is a perturbation of a normalized block basis (y_i) of the unit vector basis for $\ell_p \oplus \ell_2$. Thus if $\|y_i\|_{\ell_p} = c_i$ and $\|y_i\|_{\ell_2} = w_i$ then $\| \sum a_i y_i \| = (\sum |a_i|^p |c_i|^p)^{1/p} \vee (\sum |a_i|^2 w_i^2)^{1/2}$. From Proposition 5, $\| \sum a_i y_i \|_{(\ell_p \oplus \ell_2)_p} \geq (\sum |a_i|^p)^{1/p}$, hence

$$\left(\sum |a_i|^p\right)^{1/p} \vee \left(\sum |a_i|^2 w_i^2\right)^{1/2} \leq \left\|\sum a_i y_i\right\|_{(\ell_p \oplus \ell_2)_p} \leq 2 \left\|\sum a_i y_i\right\|$$

$$\leq 2 \left[\left(\sum |a_i|^p\right)^{1/p} \vee \left(\sum |a_i|^2 w_i^2\right)^{1/2} \right].$$

To see c)⇒b) we begin with a weakly null tree in S_X and choose a branch (x_i) satisfying the c) condition:

$$\left\|\sum a_i x_i\right\| \overset{K}{\sim} \left(\sum |a_i|^p\right)^{1/p} \vee \left(\sum |a_i|^2 |w_i|^2\right)^{1/2} .$$

Now we could first have "pruned" our tree so that each branch may be assumed to be a block basis of (h_i), by perturbations. We want to say that for some K',

$$\left\|\sum a_i x_i\right\| \overset{K'}{\sim} \left(\sum |a_i|^p\right)^{1/p} \vee \left\|\sum a_i x_i\right\|_2$$

(we have $\overset{K'}{\geq}$ by the fundamental inequality).

If this fails we can find a block basis (y_n) of (x_n),

$$y_n = \sum_{i=k_{n-1}+1}^{k_n} a_i x_i \ , \text{ with } \sum_{i=k_{n-1}+1}^{k_n} w_i^2 a_i^2 = 1$$

$$\text{and } \left(\sum_{i=k_{n-1}+1}^{k_n} |a_i|^p \right)^{1/p} \vee \|y_n\|_2 < 2^{-n}.$$

But then from the c) condition (y_n) is equivalent to the unit vector basis of ℓ_2 and from the above condition and the [20] argument, a subsequence is equivalent to the unit vector basis of ℓ_p, a contradiction.

Note that b)$\Rightarrow$d) since if (x_i) is a normalized weakly null sequence and we define $(x_\alpha)_{\alpha\in T_\infty}$ by $x_{(n_1,\dots,n_k)} = x_{n_k}$ then the branches of $(x_\alpha)_{\alpha\in T_\infty}$ coincide with the subsequences of (x_n). Note that the condition d) just says we may take the weight "w" in [15] to be "$\lim_i \|x_i\|_2$".

It remains to show d)$\Rightarrow$b) in Theorem 11 and this will complete the proof of Theorem 13. The idea is to use Burkholder's inequality using d) on nodes of a weakly null tree, following the scheme of [15] to accomplish this. That argument will obtain a branch $(x_n) = (x_{\alpha_n})$, $\alpha_n = (m_1, \dots, m_n)$ with

$$\left\|\sum a_i x_i\right\| \sim \left(\sum |a_i|^p\right)^{1/p} \vee \left(\sum w_i^2 a_i^2\right)^{1/2}$$

where $w_i \overset{C(p)}{\sim} \lim_n \|x_{(\alpha_n,n)}\|_2$ using d).

Our next goal is to show that if $X \subseteq L_p$ and X does not embed into $\ell_p \oplus \ell_2$, then X contains an isomorphic copy of $(\sum \ell_2)_p$. The idea will be to use the failure of d) to show $(\sum \ell_2)_p \hookrightarrow X$. In the [20] argument, we obtained a sequence $(x_i) \subseteq S_X$ with the x_i's becoming more and more skinny:

$$\lim_i m[|x_i| \geq \varepsilon] = 0 \quad \text{for all} \quad \varepsilon > 0$$

and then extracted an ℓ_p subsequence, of almost disjointly supported functions. Here we want to replace x_i by a sequence of skinny K-isomorphic copies of ℓ_2.

Theorem 13 ([12]). *Let $X \subseteq L_p$, $2 < p < \infty$. If X does not embed into $\ell_p \oplus \ell_2$ then $(\sum \ell_2)_p \hookrightarrow X$.*

We want to produce $X_i \subseteq X$, $X_i \overset{K}{\sim} \ell_2$ where two things happen. First, for all $\varepsilon > 0$ there exists i so that if $x \in S_{X_i}$ then $m[|x| \geq \varepsilon] < \varepsilon$. Secondly we need that X_i is not too skinny, namely each B_{X_i} is p-uniformly integrable.

Definition 9. $A \subseteq L_p$ is *p-uniformly integrable* if

$$\forall \varepsilon > 0 \, \exists \delta > 0 : \ \forall m(E) < \delta \, \forall \, z \in A, \ \int_E |z|^p < \varepsilon.$$

Lemma 2. *Assume for some K and all n there exists $(x_i^n)_{n=1}^\infty \subseteq S_X$ with $\lim_i \|x_i^n\|_2 = \varepsilon_n \downarrow 0$ and $(x_i^n)_i$ is K-equivalent to the unit vector basis of ℓ_2. Then $(\sum \ell_2)_p \hookrightarrow X$.*

Sketch of proof. Note that if $y = \sum_i a_i x_i^n$ has norm 1 then, assuming as we may that $(x_i^n)_i$ is a block basis of (h_i) and $\|x_i^n\|_2 \approx \varepsilon_n$ then

$$\|y\|_2 \approx \left(\sum a_i^2 \|x_i^n\|_2^2\right)^{1/2} \lesssim K\varepsilon_n.$$

So we have a sequence of skinny K-ℓ_2's inside of X. We would like to have if $y^n \in [(x_i^n)_i]$ then they are essentially disjointly supported so $\|\sum y^n\| \sim (\sum \|y^n\|^p)^{1/p}$, as in the [20] argument. Unlike in [20] we cannot select one y_n from each $[(x_i^n)_i]$ and pass to a subsequence. We need to fix a given $[(x_i^n)_i]$ for large n so it is skinny enough based on the earlier selections of subspaces and also so that its unit ball is p-uniformly integrable so that future selections of $[(x_i^m)_i]$ will be essentially disjoint from it. □

To achieve this we need a sublemma.

Sublemma. *Let $Y \subseteq L_p$, $2 < p < \infty$, with $Y \sim \ell_2$. There exists $Z \subseteq Y$ with S_Z p-uniformly integrable.*

Proof. This is proved in two steps. First showing a normalized martingale difference sequence (x_n) with $\{(x_n)\}$ p-uniformly integrable has $A = \{\sum a_i x_i \colon \sum a_i^2 \leq 1\}$ also p-uniformly integrable by a stopping time argument.

The general case is to use the subsequence splitting lemma to write a subsequence of an ℓ_2 basis as $x_i = y_i + z_i$ where the (y_i) are a p-uniformly integrable (perturbation of) a martingale difference sequence and the z_i's are disjointly supported and then use an averaging argument to get a block basis where the z_i's disappear. □

The subsequence splitting lemma is a nice exercise in real analysis: Given a bounded $(x_i') \subseteq L_1$ there exists a subsequence $(x_i) \subseteq (x_i')$ with $x_i = y_i + z_i$, $y_i \wedge z_i = 0$, (y_i) is uniformly integrable and the z_i's are disjointly supported.

Now we return to condition d) in Theorem 11 and recall by [15] every weakly null sequence in S_X has a subsequence (x_i) with, for some $w \in [0,1]$,

$$\left\|\sum a_i x_i\right\| \overset{D_p}{\sim} \left(\sum |a_i|^p\right)^{1/p} \vee w\left(\sum |a_i|^2\right)^{1/2}$$

and d) asserts that for some absolute C, $w \overset{C}{\sim} \lim_i \|x_i\|_2$. Now clearly we can assume that $w \geq \lim_i \|x_i\|_2$ and if d) fails we can use this to construct our ℓ_2's satisfying the lemma and thus obtain $(\sum \ell_2)_p \hookrightarrow X$.

Indeed d) fails yields that we can take a normalized block basis (y_i) of a given (x_i) failing the condition for a large C to obtain $(y_i) \overset{D_p}{\sim} \ell_2$ basis yet $\|y_i\|_2$ remains small.

So we have the dichotomy for $X \subseteq L_p$, $2 < p < \infty$. Either

- $X \hookrightarrow \ell_p \oplus \ell_2$ or
- $(\sum \ell_2)_p \hookrightarrow X$.

In the latter case, using L_p is stable we can get for all $\varepsilon > 0$, $(\sum \ell_2)_p \overset{1+\varepsilon}{\hookrightarrow} X$.

The theory of stable spaces was developed by Krivine and Maurey ([22]). X is *stable* if for all bounded $(x_n), (y_n) \subseteq X$,

$$\lim_m \lim_n \|x_n + y_m\| = \lim_n \lim_m \|x_n + y_m\|$$

provided both limits exist. They proved that if X is stable then for some p and all $\varepsilon > 0$, $\ell_p \overset{1+\varepsilon}{\hookrightarrow} X$. They also proved L_p is stable, $1 \leq p < \infty$.

We have obtained in our proof that if $X \not\hookrightarrow \ell_p \oplus \ell_2$ then for some K and all $\varepsilon > 0$ there exist $X_n \subseteq X$, $X_n \overset{K}{\sim} \ell_2$ and if $x_n \in X_n$, $\|\sum x_n\| \overset{1+\varepsilon}{\sim} (\sum \|x_n\|^p)^{1/p}$. Using L_p is stable we can choose $Y_n \subseteq X_n$, $Y_n \overset{1+\varepsilon}{\sim} \ell_2$ for all n.

In fact we can get $(\sum \ell_2)_p$ complemented in X via the next result. We note first that if $(x_i) \subseteq S_{L_p}$ is K-equivalent to the unit vector basis of ℓ_2 then, as mentioned earlier, by [32] it is $C(K)$-complemented in L_p by some projection P. Also P must have the form (true for any projection of any space onto ℓ_2)

> $Px = \sum x_i^*(x) x_i$ where (x_i^*) is biorthogonal to (x_i) and is weakly null in $L_{p'}$ $(\frac{1}{p} + \frac{1}{p'} = 1)$.

Proposition 8. *For all n let $(y_i^n)_i$ be a normalized basic sequence in L_p, $2 < p < \infty$, which is K-equivalent to the unit vector basis of ℓ_2 and so that for $y_n \in [(y_i^n)_i]$,*

$$\left\|\sum y_n\right\| \overset{K}{\sim} \left(\sum \|y_n\|^p\right)^{1/p}.$$

Then there exists subsequences $(x_i^n)_i \subseteq (y_i^n)_i$, for each n, so that $[\{x_i^n : n, i \in \mathbb{N}\}]$ is complemented in L_p.

Proof. By [32] each $[(y_i^n)_i]$ is $C(K)$-complemented in L_p via projections $P_n = \sum_m y_m^{n\,*}(x) y_m^n$. Passing to a subsequence and using a diagonal argument and perturbing we may assume there exists a blocking (H_m^n) of (h_i), in some order over all n, m, so that for all n, m, $\operatorname{supp}(y_m^{n\,*})$, $\operatorname{supp}(y_m^n) \subseteq H_m^n$.

This uses $y_m^n \xrightarrow{w} 0$ and $y_m^{n\,*} \xrightarrow{w} 0$ (in $L_{p'}$) as $m \to \infty$ for each n. Set $Py = \sum_{n,m} y_m^{n\,*}(y) y_m^n$. We show P is bounded, hence a projection onto a copy of $(\sum \ell_2)_p$.

Let $y = \sum_{n,m} y(n,m)$, $y(n,m) \in H_n^m$.

$$\|Py\| = \left\| \sum_n \sum_m y_m^{n\,*}(y(n,m)) y_m^n \right\|$$

$$\sim \left(\sum_n \left(\sum_m |y_m^{n\,*}(y(n,m))|^2 \right)^{p/2} \right)^{1/p}.$$

Now

$$\left(\sum_m |y_m^{n\,*}(y(n,m))|^2 \right)^{1/2} \sim \|P_n y(n)\| \le C(K)\|y(n)\|$$

where $y(n) = \sum_m y(n,m)$. So

$$\|Py\| \le \bar{C}(K) \left(\sum \|y_n\|^p \right)^{1/p} \le \bar{\bar{C}}(K)\|y\|. \qquad \square$$

Remark 4. The proof of Proposition 8 above is due to Schechtman (private communication). He also proved by a different much more complicated argument that the complementation result extends to $1 < p < 2$.

In [12] the proofs of all the results are also considered using Aldous' ([1]) theory of random measures. We are able to show if $(\sum \ell_2)_p \hookrightarrow X \subseteq L_p$, $2 < p < \infty$, then given $\varepsilon > 0$ there exists $(\sum Y_n)_p \overset{1+\varepsilon}{\hookrightarrow} X$, $d(Y_n, \ell_2) < 1 + \varepsilon$ and moreover: there exist disjoint sets $A_n \subseteq [0,1]$ with for all n, $y \in Y_n$, $\|y|_{A_n}\| \ge (1 - \varepsilon 2^{-n})\|y\|$ and $[Y_n : n \in \mathbb{N}]$ is $(1+\varepsilon)\ C_p^{-1}$ complemented in L_p where C_p is the norm of a symmetric normalized Gaussian random variable in L_p. This is best possible by [11].

We can also deduce the [17] result: $X \subseteq L_p$, $2 < p < \infty$, and X is a quotient of a subspace of $\ell_p \oplus \ell_2 \Rightarrow X \hookrightarrow \ell_p \oplus \ell_2$, by showing that such an X cannot contain $(\sum \ell_2)_p$.

Lemma 3. *Let W be a subspace of $\ell_p \oplus \ell_2$. Let $X = \ell_2$, let $Q\colon W \to X$ be a quotient mapping and let λ be a constant with $0 < \lambda < \|Q\|^{-1}$. For every $M > 0$ there is a finite co-dimensional subspace Y of X such that, for $w \in W$ we have*

$$\|w\| \le M,\ Q(w) \in Y,\ \|Q(w)\| = 1 \implies \|w\|_2 > \lambda.$$

Proof. Suppose otherwise. We can find a normalized block basis (x_n) in X and elements w_n of W with $\|w_n\| \le M$, $Q(w_n) = x_n$ and $\|w_n\|_2 \le \lambda$. Taking a subsequence and perturbing slightly, we may suppose that $w_n = w + w'_n$, where (w'_n) is a block basis in $\ell_p \oplus \ell_2$, satisfying $\|w'_n\| \le M$, $\|w'_n\|_2 \le \lambda$.

Since $Q(w) = \text{w-}\lim Q(w_n) = 0$, we see that $Q(w'_n) = x_n$. We may now estimate as follows using the fact that the w'_n are disjointly supported:

$$\left\|\sum_{n=1}^{N} w'_n\right\| = \left(\sum_{n=1}^{N} \|w'_n\|_p^p\right)^{1/p} \vee \left(\sum_{n=1}^{N} \|w'_n\|_2^2\right)^{1/q} \le N^{1/p}M \vee N^{1/2}\lambda.$$

Since (x_n) is a normalized block basis in $X = \ell_2$ we have

$$N^{1/2} = \left\|\sum_{n=1}^{N} x_n\right\| \le \|Q\| \left\|\sum_{n=1}^{N} w'_n\right\| \le M\|Q\|N^{1/p} \vee \lambda\|Q\|N^{1/2}.$$

Since $\lambda\|Q\| < 1$, this is impossible once N is large enough. $\square$

Proposition 9. *$(\sum \ell_2)_p$ is not a quotient of a subspace of $\ell_p \oplus \ell_2$.*

Proof. Suppose, if possible, that there exists a quotient operator

$$\ell_p \oplus \ell_2 \supseteq Z \xrightarrow{Q} X = \left(\bigoplus_{n\in\mathbb{N}} X_n\right)_p$$

where $X_n = \ell_2$ for all n. Let K be a constant such that $Q[KB_Z] \supseteq B_X$, let λ be fixed with $0 < \lambda < \|Q\|^{-1}$, choose a natural number m with $m^{1/2-1/p} > K\lambda^{-1}$, and set $M = 2Km^{1/p}$.

Applying the lemma, we find, for each n, a finite co-dimensional subspace Y_n of X_n such that

$$z \in MB_Z,\ Q(z) \in Y_n,\ \|Q(z\| = 1 \implies \|z\|_q > \lambda. \tag{2}$$

For each n, let $(e_i^{(n)})$ be a sequence in Y_n, 1-equivalent to the unit vector basis of ℓ_2. For each m-tuple $\mathbf{i} = (i_1, i_2, \ldots, i_m) \in \mathbb{N}^m$, let $z(\mathbf{i}) \in Z$ be chosen with

$$Q(z(\mathbf{i})) = e_{i_1}^{(1)} + e_{i_2}^{(2)} + \cdots + e_{i_m}^{(m)},$$

and $\|z(\mathbf{i})\| \le Km^{1/p}\ \|e_{i_1}^{(1)} + \cdots + e_{i_m}^{(m)}\| = Km^{1/p}$.

Taking subsequences in each coordinate, we may suppose that the following weak limits exist in Z

$$z(i_1, i_2, \ldots, i_{m-1}) = \text{w-}\lim_{i_m \to \infty} z(i_1, i_2, \ldots, i_m)$$

$$\vdots$$

$$z(i_1, i_2, \ldots, i_j) = \text{w-}\lim_{i_{j+1} \to \infty} z(i_1, i_2, \ldots, i_{j+1})$$

$$\vdots$$

$$z(i_1) = \text{w-}\lim_{i_2 \to \infty} z(i_1, i_2).$$

Notice that, for all j and all $i_1, i_2, \ldots, i_j$, the following hold:

$$Q(z(i_1, \ldots, i_j) = e^{(1)}_{i_1} + \cdots + e^{(j)}_{i_j}$$

$$\|z(i_1, \ldots, i_j)\| \le K m^{1/p}$$

$$\|z(i_1, \ldots, i_j) - z(i_1, \ldots, i_{j-1})\| \le 2K m^{1/p} = M.$$

Since $Q(z(i_1, \ldots, i_j) - z(i_1, \ldots, i_{j-1})) = e^{(j)}_{i_j} \in S_{Y_j}$ it must be that

$$\|z(i_1, \ldots, i_j) - z(i_1, \ldots, i_{j-1})\|_2 > \lambda, \quad \text{[by (2)]}. \tag{3}$$

We shall now choose recursively some special i_j in such a way that $\|z(i_1, \ldots, i_j)\|_2 > \lambda j^{1/2}$ for all j. Start with $i_1 = 1$; since $\|z(i_1)\| \le M$ and $Q(z(i_1)) = e^{(1)}_{i_1}$ we certainly have $\|z(i_1)\|_2 > \lambda$ by (2). Since $z(i_1, k) - z(i_1) \to 0$ weakly we can choose i_2 such that $z(i_1, i_2) - z(i_1)$ is essentially disjoint from $z(i_1)$. More precisely, because of (3), we can ensure that

$$\|z(i_1, i_2)\|_q = \|z(i_1) + (z(i_1, i_2) - z(i_1))\|_2 > (\lambda^2 + \lambda^2)^{1/2} = \lambda 2^{1/2}.$$

Continuing in this way, we can indeed choose $i_3, \ldots, i_m$ in such a way that

$$\|z(i_1, \ldots, i_j)\|_q \ge \lambda j^{1/2}.$$

However, for $j = m$ this yields

$$\lambda m^{1/2} \le K m^{1/p},$$

contradicting our initial choice of m. □

We can also obtain some asymptotic results. First we recall the relevant definitions

$$\text{cof}(X) = \{Y \subseteq X : Y \text{ is of finite co-dimension in } X\}.$$

Definition 10 ([28]). *Let $(e_i)_1^n$ be a normalized monotone basis. $(e_i) \in \{X\}_n$, the nth asymptotic structure of X, if the following holds:*

$$\begin{aligned} &\forall\, \varepsilon > 0\ \forall\ X_1 \in \operatorname{cof}(X)\ \exists\ x_1 \in S_{X_1} \\ &\qquad\qquad \forall\ X_2 \in \operatorname{cof}(X)\ \exists\ x_2 \in S_{X_2} \\ &\qquad\qquad \ldots \\ &\qquad\qquad \forall\ X_n \in \operatorname{cof}(X)\ \exists\ x_n \in S_{X_n} \\ &\qquad \textit{with } d_b((x_i)_1^n, (e_i)_1^n) < 1 + \varepsilon. \end{aligned}$$

The latter means that for some $AB < 1+\varepsilon$ for all $(a_i)_1^n \subseteq \mathbb{R}$,

$$A^{-1} \left\| \sum_1^n a_i e_i \right\| \le \left\| \sum_1^n a_i x_i \right\| \le B \left\| \sum_1^n a_i e_i \right\|,$$

i.e., $(x_i)_1^n \overset{1+\varepsilon}{\sim} (e_i)_1^n$. $d_b(\cdot)$ is the basis distance and is defined to be the minimum of such AB.

An alternate way of looking at this when X^* is separable is that $\{X\}_n$ is the smallest closed subset of $(\mathcal{M}_n, d_b(\cdot,\cdot))$ satisfying: $\forall\, \varepsilon > 0$ every weakly null tree (of length n) in S_X admits a branch $(x_i)_1^n$ with $d_b((x_i)_1^n, \{X\}_n) < 1+\varepsilon$. Here $\mathcal{M}_n$ is the set of normalized bases of length n. The metric on $\mathcal{M}_n$ is actually $\log d_b(\cdot,\cdot)$ and $\mathcal{M}_n$ is compact under this metric.

Definition 11. X is *K-asymptotic ℓ_p* if for all n and all $(e_i)_1^n \in \{X\}_n$, $(e_i)_1^n$ is K-equivalent to the unit vector basis of ℓ_p^n.

The [16,20] results yield for $X \subseteq L_p$, $2 < p < \infty$

- X is asymptotic $\ell_p \Rightarrow X \hookrightarrow \ell_p$ (since $\ell_2 \not\hookrightarrow X$)
- X is asymptotic $\ell_2 \Rightarrow X \hookrightarrow \ell_2$ (since $\ell_p \not\hookrightarrow X$).

Definition 12. X is *asymptotically $\ell_p \oplus \ell_2$* if $\exists\, K\ \forall\, n \forall\, (e_i)_1^n \in \{X\}_n\ \exists\, (w_i)_1^n$ with

$$\left\| \sum_1^n a_i e_i \right\| \overset{K}{\sim} \left(\sum_1^n |a_i|^p \right)^{1/p} \vee \left(\sum_1^n w_i^2 a_i^2 \right)^{1/2}.$$

This just says that for some K every weakly null tree of n-levels in S_X admits a branch K-equivalent to a normalized block basis of $\ell_p \oplus \ell_2$.

Proposition 10. *Let $X \subseteq L_p$, $2 < p < \infty$. X is asymptotically $\ell_p \oplus \ell_2$ iff $X \hookrightarrow \ell_p \oplus \ell_2$.*

This follows easily from our results by showing that $(\sum \ell_2)_p$ is not asymptotically $\ell_p \oplus \ell_2$.

Indeed write $(\sum \ell_2)_p = (\sum X_n)_p$, $X_n = \ell_2$ for all n, and let $(e_i^n)_{i=1}^\infty$ be the unit vector basis of X_n. Let $(N_i)_{i=1}^\infty$ be a partition of $\mathbb{N}$ into infinite subsets. Set $(x_\alpha)_{\alpha \in T_\infty}$ to be the weakly null normalized tree in $(\sum X_n)_p$ given by

$$x_{(m_1,\dots,m_k)} = e_{m_k}^i \quad \text{if} \quad k \in N_i.$$

Then each branch of this tree is equivalent to the unit vector basis of $(\sum \ell_2)_p$, in some order. Since $(\sum \ell_2)_p \not\hookrightarrow \ell_p \oplus \ell_2$ this yields $(\sum \ell_2)_p$ is not asymptotically $\ell_p \oplus \ell_2$.

Recall our goal was to characterize when $X \subseteq L_p$ $(2 < p < \infty)$ embeds into or contains isomorphically one of the small subspaces ℓ_p ℓ_2, $\ell_p \oplus \ell_2$ or $(\sum \ell_2)_p$. We have solved this except for one case which remains open.

Problem 1. *Let $X \subseteq L_p$, $p > 2$. Give an intrinsic characterization of when $X \hookrightarrow (\sum \ell_2)_p$.*

In light of the [17] $\ell_p \oplus \ell_2$ quotient result we ask the following.

Problem 2. *Let $X \subseteq L_p$ $(2 < p < \infty)$. If X is a quotient of (a subspace of) $(\sum \ell_2)_p$ does X embed into $(\sum \ell_2)_p$?*

We also note the following

Problem 3. *Characterize when a reflexive space X embeds into $(\sum F_n)_2 \oplus (\sum G_n)_p$ for some sequences (F_n), (G_n) of finite dimensional spaces.*

The difficulty here is to find a suitable norm to replace $\|\cdot\|_2$ which naturally existed when $X \subseteq L_p$, $p > 2$.

Extensive study has been made of the $\mathcal{L}_p$ spaces. X is $\mathcal{L}_p$ if for some $\lambda > \infty$ and all finite dimensional $F \subseteq X$ there exists $F \subseteq G \subseteq X$ with $d(G, \ell_p^{\dim G}) \leq \lambda$. It is known that X is $\mathcal{L}_p$ $(1 < p < \infty)$ iff X is isomorphic to a complemented subspace of L_p and X is not isomorphic to ℓ_2 (see e.g., [23] and [24]). In particular there are uncountably many such spaces ([4]) and even infinitely many which embed into $(\sum \ell_2)_p$ ([34]). Thus it seems that a deeper study of the index in [4] will be needed for further progress. However some things, which we now recall, are known.

- ([31]) *If Y is complemented in ℓ_p then Y is isomorphic to ℓ_p* (Proposition 7).
- ([19]) *If Y is a $\mathcal{L}_p$ subspace of ℓ_p then Y is isomorphic to ℓ_p.*

- ([9]) *If Y is complemented in $\ell_p \oplus \ell_2$ then Y is isomorphic to ℓ_p, ℓ_2 or $\ell_p \oplus \ell_2$.*
- ([29]) *If Y is complemented in $(\sum \ell_2)_p$ then Y is isomorphic to ℓ_p, ℓ_2, $\ell_p \oplus \ell_2$ or $(\sum \ell_2)_p$.*

We recall that X_p is the $Ł_p$ discovered by H. Rosenthal ([33]). For $p > 2$, X_p may be defined to be the subspace of $\ell_p \oplus \ell_2$ spanned by $(e_i + w_i f_i)$, where (e_i) and (f_i) are the unit vector bases of ℓ_p and ℓ_2, respectively, and where $w_i \to 0$ with $\sum w_i^{2p/p-2} = \infty$. Since $\ell_p \oplus \ell_2$ embeds into X_p, the subspaces of X_p and of $\ell_p \oplus \ell_2$ are (up to isomorphism) the same. For $1 < p < 2$ the space X_p is defined to be the dual of $X_{p'}$ where $1/p+1/p' = 1$. When restricted to $\mathcal{L}_p$-spaces, the results above lead to a dichotomy valid for $1 < p < \infty$.

Proposition 11 ([12]). *Let Y be a $Ł_p$-space $(1 < p < \infty)$. Either Y is isomorphic to a complemented subspace of X_p or Y has a complemented subspace isomorphic to $(\sum \ell_2)_p$.*

Proof. For $p > 2$ it is shown in [17] that a $Ł_p$-space which embeds in $\ell_p \oplus \ell_2$ embeds complementedly in X_p. Combining this with the main theorem of the present paper gives what we want for $p > 2$. When $1 < p < 2$, a simple duality argument extends the result to the full range $1 < p < \infty$. □

It remains a challenging problem to understand more deeply the structure of the $Ł_p$-subspaces of X_p and $\ell_p \oplus \ell_2$. If X has an unconditional basis we know the answer.

- ([17]) *If Y is a $Ł_p$ subspace of $\ell_p \oplus \ell_2$ (or X_p), $2 < p < \infty$, and Y has an unconditional basis then Y is isomorphic to ℓ_p, $\ell_p \oplus \ell_2$ or X_p.*
- ([17]) *If Y is a $Ł_p$ subspace of $\ell_p \oplus \ell_2$ $(1 < p < 2)$ with an unconditional basis then Y is isomorphic to ℓ_p or $\ell_p \oplus \ell_2$.*

It is known ([18]) that every $Ł_p$ space has a basis but it remains open if it has an unconditional basis.

So the main open problem for small $Ł_p$ spaces is to overcome the unconditional basis obstacle.

Problem 4.

(a) *Let X be a $Ł_p$ subspace of $\ell_p \oplus \ell_2$ $(2 < p < \infty)$. Is X isomorphic to ℓ_p, $\ell_p \oplus \ell_2$ or X_p?*

(b) Let X be a L_p subspace of $\ell_p \oplus \ell_2$ $(1 < p < \infty)$. Is X isomorphic to ℓ_p or $\ell_p \oplus \ell_2$?

References

1. D. Aldous, *Subspaces of L^1, via random measures*, Trans. Amer. Math. Soc. **267**(2) (1981), 445–463.
2. D. Alspach and E. Odell, *L_p spaces*, Handbook of the geometry of Banach spaces, vol. I, 123–159, North-Holland, Amsterdam, 2001.
3. F. Albiac and N.J. Kalton, *Topics in Banach space theory*, Graduate Texts in Mathematics, 233, Springer, New York, 2006.
4. J. Bourgain, H. P. Rosenthal and G. Schechtman, *An ordinal L^p-index for Banach spaces, with application to complemented subspaces of L^p*, Ann. of Math. **114**(2) (1981), 193–228.
5. D. L. Burkholder, *Martingales and singular integrals in Banach spaces*, Handbook of the geometry of Banach spaces, vol.I, 233–269, North–Holland, Amsterdam, 2001.
6. D. L. Burkholder, *Distribution function inequalities for martingales*, Ann. Probability **1** (1973), 19–42.
7. D. L. Burkholder, B. J. Davis and R. F. Gundy, *Integral inequalities for convex functions of operators on martingales*, Proc. of the 6-th Berkeley Symp. on Math. Stat. and Probl. vol. 2 (1972), 223–240.
8. J. Diestel, *Sequences and Series in Banach spaces*, Graduate Texts in Mathematics, Springer-Verlag, 1984.
9. I.S. Edelstein and P. Wojtaszczyk, *On projections and unconditional bases in direct sums of Banach spaces*, Studia Math. **56**(3) (1976), 263–276.
10. M. Fabian, P. Habala, P. Hájek, V. Montesinos, J. Pelant and V. Zizler, *Functional analysis and infinite–dimensional geometry*, CMS Books in Mathematics, Springer-Verlag, New York, 2001, x+451pp.
11. Y. Gordon, D. R. Lewis and J. R. Retherford, *Banach ideals of operators with applications*, J. Functional Analysis **14** (1973), 85–129.
12. R. Haydon, E. Odell and T. Schlumprecht, *Small subspaces of L_p*, preprint.
13. W. B. Johnson, *On quotients of L_p which are quotients of ℓ_p*, Compositio Math. **34** (1977), 69–89.
14. W. B. Johnson, B. Maurey and G. Schechtman, *Weakly null sequences in L_1*, J. Amer. Math. Soc. **20** (2007), 25–36.
15. W. B. Johnson, B. Maurey, G. Schechtman and L. Tzafriri, *Symmetric structures in Banach spaces*, Mem. Amer. Math. Soc. **19**(217) (1979), v+298.
16. W. B. Johnson and E. Odell, *Subspaces of L_p which embed into ℓ_p*, Compositio Math. **28** (1974), 37–49.
17. W. B. Johnson and E. Odell, *Subspaces and quotients of $\ell_p \oplus \ell_2$ and X_p*, Acta Math. **147** (1981), 117–147.
18. W. B. Johnson, H. P. Rosenthal and M. Zippin, *On bases, finite dimensional decompositions and weaker structures in Banach spaces*, Israel J. Math. **9** (1971), 488–506.

19. W. B. Johnson and M. Zippin, *On subspaces of quotients of* $(\sum G_n)_{\ell_p}$ *and* $(\sum G_n)_{c_0}$, Israel J. Math. **13** (1972), 311–316.
20. M. I. Kadets and A. Pełczyński, *Bases lacunary sequences and complemented subspaces in the spaces* L_p, Studia Math. **21** (1961/1962), 161–176.
21. N. J. Kalton and D. Werner, *Property* (M), M*-ideals, and almost isometric structure of Banach spaces*, J. Reine Angew. Math. **461** (1995), 137–178.
22. J. L. Krivine and B. Maurey, *Espaces de Banach stables*, Israel J. Math. **39**(4) (1981), 273–295.
23. J. Lindenstrauss and A. Pełczyński, *Absolutely summing operators in* $\mathcal{L}_p$ *spaces and their applications*, Studia Math. **29** (1968), 275–321.
24. J. Lindenstrauss and H. P. Rosenthal, *The* $\mathcal{L}_p$ *spaces*, Israel J. Math. **7** (1969), 325–349.
25. J. Lindenstrauss and L. Tzafriri, *Classical Banach Spaces I*, Springer-Verlag, New York, 1977.
26. J. Lindenstrauss and L. Tzafriri, *On the complemented subspaces problem*, Israel J. Math. **9** (1971), 263–269.
27. D. A. Martin, *Borel determinancy*, Annals of Math. **102** (1975), 363–371.
28. B. Maurey, V. D. Milman, and N. Tomzczak-Jaegermann, *Asymptotic infinite dimensional theory of Banach spaces*, Oper. Theory: Adv. Appl. **77** (1994), 149–175.
29. E. Odell, *On complemented subspaces of* $(\sum \ell_2)_{\ell_p}$, Israel J. Math. **23**(3-4) (1976), 353–367.
30. E. Odell and T. Schlumprecht, *Trees and branches in Banach spaces*, Trans. Am. Math. Soc. **354** (2002), 4085–4108.
31. A. Pełczyński, *Projections in certain Banach spaces*, Studia Math. **19** (1960), 209–228.
32. A. Pełczyński and H. Rosenthal, *Localization techniques in* L^p *spaces*, Studia Math. **52** (1974/75), 263–289.
33. H. P. Rosenthal, *On the subspaces of* L^p $(p > 2)$ *spanned by sequences of independent random variables*, Israel J. Math. **8** (1970), 273–303.
34. G. Schechtman, *Examples of* $\mathcal{L}_p$ *spaces* $(1 < p \neq 2 < \infty)$, Israel J. Math. **22** (1975), 138–147.
35. G. Schechtman, *A remark on unconditional basic sequences in* L_p $(1 < p < \infty)$, Israel J. Math. **19** (1974), 220–224.

PROBLEMS ON HYPERCYCLIC OPERATORS

HÉCTOR N. SALAS

Department of Mathematics
Universidad de Puerto Rico
Mayagüez, Puerto Rico 00681
E-mail: salas@math.uprm.edu

In this talk we review properties of hypercyclic operators, revisit some well known results, and present several problems, some of them in the new classes of dual hypercyclic operators and frequently hypercyclic operators.

1. Continuous functions with dense orbits

When studying a continuous function acting on a topological space, it is often convenient to look at the orbit of elements and their closures. The two extremes for these orbits would be either finite, in which case the element is periodic, or the whole space. If the space is metric and perfect, then the second alternative is not possible, but perhaps the closure of the orbit is the whole space. We are interested in this kind of phenomena with the additional requirements that the space has a linear structure and the function is linear. But first let's look at a very simple situation. Let $\mathbb{T} = \{e^{2\pi it} \colon t \in \mathbb{R}\}$ be the unit circle in the complex plane $\mathbb{C}$, and let f be the function defined on $\mathbb{T}$ by $f(z) = e^{i2\pi\theta}z$. The action of f is very different according to θ being rational or irrational. In the first case each point of $\mathbb{T}$ is periodic, whereas in the second case each point has a dense orbit. This result was known by Dirichlet in 1845. The function $g(z) = z^2$ has a more complicated behavior. There are some points for which the orbit of z under g is periodic, others for which it is dense, and still others for which both the preceding statements are not true. Our last example for now is Kroenecker's theorem, chapter 23 of [23], which says that in the n-torus $\mathbb{T}^n$ the function $f((z_1, \cdots, z_2)) = (e^{2\pi i\theta_1}z_1, \cdots, e^{2\pi i\theta_n}z_n)$ has dense orbit whenever $\{1, \theta_1, \cdots, \theta_n\}$ is a linearly independent set over $\mathbb{Q}$.

Let F be a Fréchet space and let $\mathcal{L}(F)$ denote the continuous linear operators on F. An operator $T \in \mathcal{L}(F)$ is *hypercyclic* if there exists an

$x \in F$ such that

$$\text{Orb}\,(T,x) := \{T^n x\colon n = 0,1,2,\ldots\}$$

is dense in F, in which case F has to be infinite–dimensional and separable. Let $\mathcal{H}_F$ denote the hypercyclic operators on F and let

$$\mathcal{H}(T) := \{x \in F\colon \overline{\text{Orb}\,(T,x)} = F\}.$$

The three following examples have become classic. Notice how they are separated in time.

(1) Birkhoff ([7]) showed in 1929 that translation operators on Hol ($\mathbb{C}$) endowed with the compact–open topology are hypercyclic.
(2) In 1952, MacLane ([29]) did the same for the differentiation operator D on Hol ($\mathbb{C}$).
(3) The first example on Banach spaces was given by Rolewicz in 1967 ([33]). For $1 \le p < \infty$, let $S \in \mathcal{L}(\ell^p(\mathbb{N}))$ be the backward shift $S(e_0) = 0$ and $S(e_n) = e_{n-1}$, where $\{e_n\colon n \in \mathbb{N}\}$ is the canonical basis. Then $\lambda S \in \mathcal{H}_{\ell^p(\mathbb{N})}$ whenever $|\lambda| > 1$.

The subject was dormant for several years. In 1982 Kitai ([25]) discovered several key properties of hypercyclicity. Unfortunately, the only portion of her thesis that she published was the last chapter ([26]). Hypercyclicity has a very rich history and has contact with several other areas. In the 1999 survey [21], Grosse-Erdmann gives a wealth of information. A very nice treatment of hypercyclicty is given by Shapiro in [37]. Our bibliography is only a very small portion of the published literature in this area.

An important tool for proving hypercylicity is the Hypercyclicity Criterion. It was proposed by Kitai ([25]) and also by Gethner and Shapiro ([17]) for the full sequence $(n)_n$. It is used in the examples 5-7 below.

Theorem 1.1. *Let F be a Fréchet space and $T \in \mathcal{L}(F)$. If T satisfies the Hypercyclicity Criterion with respect to the subsequence $(n_k)_k$ provided there exist dense subsets X and Y of F and (possibly discontinuous) mappings $S_k\colon Y \longrightarrow F$ $(k = 1,2,\ldots)$ so that*

(1) $T^{n_k} \longrightarrow 0$ pointwise on X,
(2) $S_{n_k} \longrightarrow 0$ pointwise on Y,
(3) $T^{n_k} S_{n_k} \longrightarrow I_Y$ (Identity on Y).

Then T is hypercyclic.

In [17] this criterion was used to give a unified proof for the classical examples. Let's use it for example 2 as is done in [37]. Let $X = Y$ be the space of polynomials in Hol $(\mathbb{C})$. Let D denote the differentiation operator, and let S denote the operator of integration from 0 to z. Then for each polynomial p is true that $D^n p$ is eventually zero, $S^n p \to 0$ uniformly on compact subsets of $\mathbb{C}$, and $DSp = p$. Thus the hypotheses of the criterion are satisfied and so D is hypercyclic.

More examples of hypercyclic operators:

(4) There are some weighted shifts, unilateral backward and bilateral; also all identity plus unilateral backward shifts are hypercyclic ([17,18,34]).
(5) Recall that if φ is a self map of the unit disc $U = \{z \in \mathbb{C}\colon |z| < 1\}$, the composition operator $C_\varphi(f) = C_\varphi \circ f$ is bounded on the Hardy space of the unit disc, $H^2(U)$. Bourdon and Shapiro ([13]) initiated the study of which composition operators are hypercyclic.
(6) Godefroy and Shapiro ([18]) showed that some adjoint of multipliers are hypercyclic.
(7) Herrero ([24]) discovered the surprising fact that some compact perturbations of the identity are in $\mathcal{H}_F$.
(8) An easy consequence of Theorem 1.1 is: If T_j satisfy the Hypercylicity Criterion for the same sequence, then $\oplus_{j=1}^{\infty} T_j \in \mathcal{H}_{\oplus_{j=1}^{\infty} B_j}$.
(9) There are operators in $\mathcal{H}_B$ that do not satisfy the Hypercylicity Criterion. This is a recent work by De la Rosa and Read ([15]).

2. Some properties of hypercyclic operators

F will denote a a Fréchet space, and B a Banach space. If $T \in \mathcal{H}_F$, then:

(1) $S^{-1}TS \in \mathcal{H}_F$, whenever S is invertible. Thus the set $\mathcal{H}_F$ is invariant under similarity.
(2) Suppose that $F = E \oplus G$ i.e., the projection P on E along G is continuous (E is the fixed subspace of P whereas $G =$ Ker (P)). Suppose further that $T(G) \subset G$; i.e, G is invariant under T, then $PTP \in \mathcal{H}_E$. This was first observed by Herrero when F is a Hilbert space, but the same idea works in general. Indeed, let $y \in E$ and let $x = Px \oplus (I - P)x \in \mathcal{H}_F(T)$. Therefore there is a subsequence $T^{n_k}(x) \to y$ and since $T^n = (PTP)^n \oplus (I - P)S_n$ it follows that $(PTP)^{n_k}(Px) \to y$. Furthermore, if E and G are both invariant, then $T|_E \in \mathcal{H}_E$ and $T|_G \in \mathcal{H}_G$.
(3) It was shown in [25] that $\sigma(T)$ meets the unit circle $\mathbb{T}$ and moreover, every component of $\sigma(T)$ must meet $\mathbb{T}$. To see the last part, let σ_1 and

σ_2 be a decomposition of $\sigma(T)$ in two disjoint and nonempty closed sets. Their corresponding Riesz projections decompose $F = F_1 \oplus F_2$ and $\sigma(T|_{F_j}) = \sigma_j$ for $j = 1, 2$. By the last observation in the second property, it follows that σ_j meets the unit circle.

(4) $\mathcal{H}(T)$ is a G_δ dense set ([17]).
(5) Moreover, in [9] Bourdon showed that: If x is hypercyclic, then $\{P(T)x\colon P \text{ is a nonzero polynomial}\}$ is a connected subset of $\mathcal{H}(T)$.
(6) The point spectrum $\sigma_p(T^*)$ is empty.
(7) $T \in \mathcal{H}_F$ if and only if T is *topologically transitive*; i.e., given non empty open subsets U and V of F there exists n such that $T^n(V) \cap U$ is non empty ([18]). A stronger property is *mixing*; i.e., given non empty open subsets U and V of F there exists n_0 such that $T^n(V) \cap U$ is non empty for all $n \geq n_0$ (see Costakis and Sambarino [14]).
(8) In [1], Ansari showed that:

Theorem 2.1. *Let B be a Banach space and let $T \in \mathcal{L}(B)$. If T is hypercyclic, then so is T^n for all $n \in \mathbb{N}$ and $\mathcal{H}(T) = \mathcal{H}(T^n)$.*

There are two immediate consequences.

Corollary 2.1. *If $T \in \mathcal{H}_B$ and $\lambda = e^{2(m/n)\pi i}$, then $\lambda T \in \mathcal{H}_B$.*

Corollary 2.2. *If $T \in \mathcal{H}_B$ and $\sigma(T^n) \neq \sigma(T^m)$ whenever $n \neq m$, then Ansari's theorem generates infinitely many hypercyclic operators none of them similar to any other.*

In [30], a new proof of this theorem is given. We give an outline of this proof in the next section.

(9) In [12], Bourdon and Feldman proved a more general result:

Theorem 2.2. *Let $T \in \mathcal{L}(F)$. If $\overline{\text{Orb}\,(T, x)}$ has non empty interior, then $T \in \mathcal{H}(F)$.*

(10) In view of Corollary 2.1, Bès asked if this is also the case for $|\lambda| = 1$. In [27], León-Saavedra and Müller gave an affirmative answer.

Theorem 2.3. *If $T \in \mathcal{H}_B$, then $\lambda T \in \mathcal{H}_B$ and $\mathcal{H}(T) = \mathcal{H}(\lambda T)$ whenever $\lambda \in \mathbb{T}$.*

In Section 4, we indicate a possible non linear extension of this result which was also obtained in [30].

(11) If B is an infinite-dimensional separable Banach space, then $\mathcal{H}_B$ is non empty. This is an old question of Rolewicz ([33]) which was answered affirmatively by Ansari ([2]) and also by Bernal-González ([5]), and for the general Fréchet case by Bonet and Peris ([8]).

(12) Moreover, every X countable dense subset of B is contained in the orbit of some T. When the space is a Hilbert space, it was proved by Halperin, Kitai and Rosenthal ([26]). Their construction is quite intricate, uses the geometry of Hilbert spaces in a fundamental way and is obtained from first principles. It seems difficult to generalize this proof to the general case when the space is Banach. However, in [20] Grivaux proved that:

Theorem 2.4. *Let Y be a countable dense linearly independent subset of the Banach space B. Then there exists $T = I + K \in \mathcal{H}_B$ such that* Orb $\{T, x\} = Y$, *where K is compact and $\|K\| < \epsilon$.*

Proof. First she proves a proposition showing that if Y and V are two countable dense sets and $\gamma > 0$, then there exists an isomorphism $L \in \mathcal{L}(B)$ such that $L(V) = Y$ with $\|I - L\| < \gamma$. The next step is to choose $T_0 = I + K_0 \in \mathcal{H}_B$ with K_0 compact and $\|K_0\| < \epsilon$, which can be done by [2] and [5]. Let $x_0 \in \mathcal{H}(T_0)$ and let $V =$ Orb $\{T_0, x_0\}$. Set $y_0 = L(x_0)$ and $T = LT_0L^{-1}$. Then $\{T^n(y_0) = L(T_0^n x_0) : n = 0, 1, 2, \ldots\} = Y$. By choosing γ to be arbitrarily small, the theorem is proved. □

Problem 2.1. *Let H be an infinite–dimensional separable Hilbert space. Let σ be a compact set of $\mathbb{C}$ such that every component intersects $\mathbb{T}$. Must there exist $T \in \mathcal{H}_H$ such that $\sigma = \sigma(T)$? More generally, for each separable Banach space B, characterize those σ for which there exists $T \in \mathcal{H}_B$ with $\sigma = \sigma(T)$.*

Recall that there are Banach spaces with few operators, for instance. if they are H.I. (hereditarily indecomposable, [19]). Thus the underlying space determines how big the spectrum may be.

Example of operators which are never hypercyclic:

(1) Contractions, since if $\|T\| \leq 1$, then Orb (T, x) is trapped in $\{\|y\| \leq \|x\|\}$.
(2) A finite dimensional operator T. The fastest proof is that $\sigma(T^*)$ is non empty ([37]).
(3) Compact operators, by spectral reasons.
(4) Identity plus finite rank operators.

(5) Normal operators on Hilbert spaces ([25]).

An operator T is said to be *supercyclic* if there is a vector $x \in F$ such that $\{\lambda T^n x \colon \lambda \in \mathbb{C} \text{ and } n = 0, 1, 2, \ldots\}$ is dense in F. The following operators are not even supercyclic:

(6) Hyponormal operators. This was proved by Bourdon ([11]). (An operator T in a Hilbert space is hyponormal if $T^*T - TT^* \geq 0$.)
(7) Surjective isometries on Banach spaces. This was proved by Ansari and Bourdon ([3]).

3. Ansari's theorem

The results in this section are from [30]. The separation theorem below is also valid for topological spaces which are not Hausdorff. For $f \colon X \longrightarrow X$ we use the notation $f^n = f \circ \cdots \circ f$; i.e., f is composed with itself n times. In [10], Bourdon proved a special case of Theorem 2.1; namely, that T^2 is hypercyclic. This is the approach that we follow here. But this approach doesn't seem to yield the more general Theorem 2.2.

Theorem 3.1. *Let (X, ρ) be a metric space without isolated points, and let f be a continuous function from X into itself. Suppose that there exists $x \in X$ with* Orb (f, x) *dense in X but* Orb (f^p, x) *is not dense, where p is a prime number.*

Then there exist p nonempty open sets $G_1, \cdots, G_p$ separating the set

$$D = \{z \in X \colon \overline{\text{Orb}\,(f, z)} = X\}$$

and $f(G_k \cap D) \subset G_{k+1} \cap D$ for $k < p$ and $f(G_p \cap D) \subset G_1 \cap D$.

Proof. We will just sketch the proof. For $1 \leq k \leq p$, set

$$A_k = \overline{\text{Orb}\,(f^p, f^{(k-1)}x)}.$$

By hypothesis, A_1 is not X. Let P be the set $\{1, 2, \ldots, p\}$. The open sets G_k are defined by

$$G_k = X \setminus \bigcup_{P \setminus \{k\}} A_j.$$

Since $\bigcup_{j \in P} A_j = X$, these sets are pairwise disjoint: for $j \neq k$

$$G_k \cap G_j = X \setminus \bigcup_{j \in P} A_j = \emptyset.$$

The following assertion is used to prove that each G_k is not empty.

Assertion: $f^{j-1}x \notin A_k$ whenever $1 \leq j \leq p$ and $j \neq k$.

If $f^{j-1}x \in A_k$, then there would be an $1 < s \leq p$ such that $A_s \subset A_1$. But then, since p is prime it would follow that $A_t \subset A_1$ for $t = 2, \ldots, p$, but this is impossible because A_1 is not X.

To see that $D \subset \bigcup_{j \in P} G_j$ is enough to show that, for $j \neq k$,

$$D \cap A_j \cap A_k = \emptyset.$$

If $z \in D$, then $\text{Orb}\,(f, z) \subset D$ since X does not have isolated points; in particular $fz \in D$. On the other hand

$$f(A_k) \subset A_{k+1} \quad \text{for} \quad 1 \leq k < p \quad \text{and } f(A_p) \subset A_1.$$

Consequently, if $z \in G_k$, then $fz \in G_{k+1}$ when $k < p$ and $fz \in G_1$ when $k = p$. □

Corollary 3.1. *Let (X, ρ) be a metric space without isolated points and let f be a continuous function from X into itself. Suppose that there exists $x \in X$ with* Orb (f, x) *dense in X. Suppose further that*

$$\text{Orb}\,(f, x) \subset E \subset D = \{z \in X \colon \overline{\text{Orb}\,(f, z)} = X\}$$

with E connected. Then Orb $(f^n, f^m x)$ *is dense in X for any pair n, m.*

As a consequence of the above theorem and its corollary, we recover Theorem 2.1. The Banach space can be over the complex numbers or the real numbers (that the space can be Fréchet and the scalars can be the real numbers is because of [6]).

Proof. (*of Ansari's theorem*) When p is prime, we use the separation theorem above and argue as Bourdon's Theorem 3.4 of [10] since

$$\{P(T)x \colon P \text{ is a non zero polynomial}\}$$

is a connected subset of $\mathcal{H}(T)$ ([9]). We prove the general case by induction. Assume that T^q is hypercyclic whenever $q < n$. The only case that is necessary to consider is when n is not prime. In that case, write $n = mp$ with p prime. Since $m < n$ we have that T^m is hypercyclic, but then $T^n = (T^m)^p$ is also hypercyclic because p is prime. Moreover, Corollary 3.1 implies that $\mathcal{H}(T) = \mathcal{H}(T^n)$. □

4. Variations on a theme of Léon–Saavedra and Müller

The results in this section are also taken from [30]. Theorem 2.3 is true also in the context of composition operators C_φ on Hol (U) where φ is an holomorphic self map of the open unit disk U. Shapiro showed in [37] that C_φ is hypercyclic if and only if φ is univalent and doesn't have a fixed point on U. Moreover, he showed that these operators are chaotic, which means that they have in addition a dense set of periodic points. In [40], Yousefi and Rezaei studied hypercyclicity for weighted composition operators $M_f C_\varphi$ on $H(U)$, where $f \in H(U)$. In particular, they showed that when $|\lambda| = 1$ and C_φ is hypercyclic, so is λC_φ. These results motivate the following:

Problem 4.1. *Suppose that X is a topological space such that, for $\lambda \in \mathbb{T}$ and $x \in X$, the multiplication $M_\lambda x = \lambda x$ makes sense and $M_\lambda \colon X \longrightarrow X$ is continuous. Let $f \colon X \longrightarrow X$ be a map with a dense orbit. What are the conditions on X and f for which λf has also a dense orbit for all $\lambda \in \mathbb{T}$?*

An example of such an X is when it is a subset of a topological vector space over the complex numbers and such that it is invariant under M_λ for all $\lambda \in \mathbb{T}$.

Another example is the regular torus since it can be seen as $\mathbb{T} \times \mathbb{T}$; i.e., the 2-torus. More generally, we can consider the n-th torus and the infinite torus

$$\mathbb{T}^n = \underbrace{\mathbb{T} \times \cdots \times \mathbb{T}}_{n} \qquad \text{and} \qquad \mathbb{T}^\infty = \prod_{j=1}^{\infty} T_j,$$

with $T_j = \mathbb{T}$ for all $j \in \mathbb{N}$ and $\mathbb{T}^\infty$ is equipped with the product topology.

$\mathbb{T}^\infty$ is an abelian topological group with multiplication coordinate–wise and with normalized Haar measure.

An easy example (which we have already mentioned in the introduction) showing that the answer to the question is not always positive is the map $f \colon \mathbb{T} \longrightarrow \mathbb{T}$ defined by $f(z) = \alpha z$, where $\alpha = e^{2\pi t i}$ is such that $t \in \mathbb{R} \setminus \mathbb{Q}$. More generally we have:

Proposition 4.1. *For each $j \in \mathbb{N}$, let $\alpha_j = e^{2\pi t_j i}$ be such that the sets $\{1, t_1, \ldots, t_n\}$ are linearly independent over $\mathbb{Q}$ for all $n \in \mathbb{N}$. Let $f \colon \mathbb{T}^\infty \longrightarrow \mathbb{T}^\infty$ be defined as $f(z_1, z_2, \ldots) = (\alpha_1 z_1, \alpha_2 z_2, \ldots)$. Then f has a dense orbit but $\alpha_j^{-1} f$ doesn't have a dense orbit for all $j \in \mathbb{N}$.*

Proof. We claim that the point $(1, 1, \ldots)$ has a dense orbit under f. Let $(z_1, z_2, \ldots)$ be a point of $\mathbb{T}^\infty$. Let $\epsilon > 0$. Since $\mathbb{T}^\infty$ has the product topology, it is enough to find an n such that $|\alpha_j^n - z_j| < \epsilon$ for $j = 1, 2, \ldots, m$. But this is what Kroenecker's theorem says, p. 381 in [23]. □

Let $(X, \mathcal{M}, \mu)$ be a probability space and let V be a measure–preserving map on X, i.e., $\mu(V^{-1}(A)) = \mu(A)$ for every measurable set $A \in \mathcal{M}$. Recall that V is called *ergodic* if whenever $A \in \mathcal{M}$ is invariant under V then $\mu(A)$ is either 0 or 1. According to Theorem 1.5 of [39], this is equivalent to saying that whenever A and B are measurable sets with positive measure, there is an $n \in \mathbb{N}$ such that $\mu(V^{-n}(A) \cap B) > 0$. Thus ergodicity is a stronger property than topological transitivity. However, one of the consequences of Theorem 1.11 of [39] is that if X is a connected, metric, compact abelian group and μ is its normalized Haar measure and B is a continuous epimorphism, then B being ergodic is equivalent to being topologically transitive, in which case the set of points with dense orbit has measure 1. It is well–known that the only endomorphisms of the unit circle $\mathbb{T}$ are of the form $\varphi(z) = z^n$ with $n \in \mathbb{Z}$. They are ergodic, and therefore topologically transitive, for $|n| > 1$.

The map in the proposition above is ergodic with respect to Haar measure in $\mathbb{T}^\infty$. More related results are obtained in [30].

5. Dual hypercyclic operators

In [34], it was given a bilateral weighted shift T on $\ell^2(\mathbb{Z})$ such that T and T^* are hypercyclic. It happens that $T \oplus T^*$ is not even cyclic. This is a consequence of the weights of T being positive and the unpublished 1982 theorem of Deddens below:

Theorem 5.1. *Let H be a separable Hilbert space. Suppose $T \in \mathcal{L}(H)$ and its matrix with respect to some orthonormal basis consists only of real entries. Then $T \oplus T^*$ is not cyclic.*

In his note [37], Shapiro gave a proof of this result. We should notice that in this case T^* is the Hilbert space adjoint which is different of the adjoint in the Banach space sense when the scalar field is $\mathbb{C}$.

An operator $T \in \mathcal{H}_B$ is called *dual hypercyclic* if T and T^* are hypercyclic. In particular this means that B^* must be separable. In [32], Petersson showed:

Theorem 5.2. *Any Banach spaces with shrinking symmetric Schauder basis support dual hypercyclic operators.*

These operators were basically bilateral weighted shifts. He asked which separable Banach spaces with separable duals support dual hypercylic operators.

We can present the following curious proposition whose proof is similar but even simpler than the proof of Deddens' theorem.

Proposition 5.1. *Let B be a separable Banach space. If $T \in \mathcal{L}(B)$, then $T \oplus T^* \in \mathcal{L}(B \oplus B^*)$ is not cyclic.*

Proof. Let $x \oplus \varphi \in B \oplus B^*$, where $\|x \oplus \varphi\| = \|x\|_B + \|\varphi\|_{B^*}$. By Hahn–Banach's theorem, it suffices to find a continuous linear function g on $B \oplus B^*$ such that $g((T \oplus T^*)^n(x \oplus \varphi) = 0$ for $n = 0, 1, 2, \ldots$. Choose $g = -\varphi \oplus i(x)$ where i is the canonical injection of B into B^{**}. □

We can extend the conclusion in [37] to dual hypercyclic operators:

Corollary 5.1. *The operators T and T^* cannot be both mixing.*

The following material until the mentioning of Volterra operators is taken from [36].

Recall that a Schauder basis $\{e_n : n \in \mathbb{Z}\}$ of a Banach space B is uncondicional if and only if $\{e_{\pi(n)} : n \in \mathbb{Z}\}$ also forms a basis for any permutation π of $\mathbb{Z}$. It is a symmetric basis if, in addition, all $\{e_{\pi(n)} : n \in \mathbb{Z}\}$, where π is a permutation, are equivalent.

Proposition 5.2. *Let $\{e_n : n \in \mathbb{Z}\}$ be a symmetric basis of B with corresponding biorthogonal functionals $\{e_n^* : n \in \mathbb{Z}\}$ and let $\{w_n : n \in \mathbb{Z}\}$ be a positive bounded sequence. If $T = \sum_{n \in \mathbb{Z}} w_n e_n^* \otimes e_{n-1}$, then $\sigma(T)$ has circular symmetry.*

It is well–known ([38]) that bilateral weighted shifts on $\ell^p(\mathbb{Z})$, with $1 \leq p < \infty$, have spectra that could be the origin, a disk, a circle or an annulus, all of them centered at the origin.

Problem 5.1. *Do the spectra of these "bilateral weighted shifts" also look like those four sets?*

The system $\{(x_n, x_n^*) : x_n \in B, x_n^* \in B^*, n \in \mathbb{Z}\}$ is called biorthogonal if $x_n^*(x_m) = \delta_n^m$. If, in addition, $[x_n : n \in \mathbb{Z}] = B$ and B^* is the weak* closure of the linear span of $\{x_n^* : n \in \mathbb{Z}\}$, then $\{x_n : n \in \mathbb{Z}\}$ is called a Markushevich basis. Note that the (biorthogonal) functionals $\{x_n^* : n \in \mathbb{Z}\}$ are unique for such a basis $\{x_n : n \in \mathbb{Z}\}$.

Ovsepian and Pelczyński showed (see p. 44 of [28]) that every separable Banach space has a Markushevich basis in which

$$\|x_n\| = 1 \text{ for all } n \text{ and } \sup_n \|x_n^*\| < \infty.$$

We call this property (OP). Moreover, when B^* is separable, $\{x_n : n \in \mathbb{Z}\}$ may be chosen so that $[x_n^* : n \in \mathbb{Z}] = B^*$. When B^* is separable, *we will always consider that property* (OP) *includes* $[x_n^* : n \in \mathbb{Z}] = B^*$. Recall that for $y \in B$ and $y^* \in B^*$, the tensor product $y^* \otimes y \in \mathcal{L}(B)$ is defined by $y^* \otimes y(x) = y^*(x)y$ and $\|y^* \otimes y\| \leq \|y^*\| \, \|y\|$. Also $(y^* \otimes y)^* = y \otimes y^*$ if we identify y with $i(y)$ where $i : B \longrightarrow B^{**}$ is the canonical injection.

Proposition 5.3. *Let B be a Banach space with Markushevich basis $\{x_n : n \in \mathbb{Z}\}$ which satisfies (OP). Let $w_n > 0$ for all n and $\sum_{n\in\mathbb{Z}} w_n < \infty$. Then the "bilateral weighted shift" $T = \sum_{n\in\mathbb{Z}} w_n x_n^* \otimes x_{n-1}$ is compact and quasinilpotent. Its adjoint T^* is $\sum_{n\in\mathbb{Z}} w_n x_{n-1} \otimes x_n^*$.*

If a "shift" T is hypercyclic, then the circular symmetry of $\sigma(T)$ and Kitai's result imply that $\sigma(T)$ must be a disk centered at the origin containing $\mathbb{T}$ or an annulus centered at the origin containing $\mathbb{T}$, or just $\mathbb{T}$.

Observe that in the theorem below, T is compact. Also there is an H.I. space constructed by Gowers and Maurey which is reflexive ([19]). But even this space supports dual hypercyclic operators.

Theorem 5.3. *Suppose that B is a Banach space whose dual is separable and with Markushevich basis $\{e_n : n \in \mathbb{Z}\}$ which satisfies (OP). Let $w_n > 0$ for all n and $\sum_{n\in\mathbb{Z}} w_n < \infty$. Then there exists a dual hypercyclic operator*

$$I + \sum_{n\in\mathbb{Z}} v_n e_n^* \otimes e_{n-1} = I + T$$

such that all $v_n > 0$ and $v_n = w_n$ except, possibly, for $v_0 \leq w_0, v_4 \leq w_4, v_{-4} \leq w_{-4}, v_{12} \leq w_{12}$ and, for $k > 2$,

$$v_{-(4+\sum_{i=2}^{k-1} 2^{2i})} \leq w_{-(4+\sum_{i=2}^{k-1} 2^{2i})}$$

and

$$v_{4+\sum_{i=2}^{k} 2^{2i-1}} \leq w_{4+\sum_{i=2}^{k} 2^{2i-1}}.$$

In the next theorem, V may be non-quasinilpotent and therefore $\sigma(I + V)$ may be a disk centered in 1 with positive radius. Thus $\sigma((I + V)^n)$ are more interesting.

Theorem 5.4. *Let L be a bilateral weighted shift on $\ell^2(\mathbb{Z})$, with positive bounded weight sequence $\{w_n\}$. Then there exists another bilateral shift V*

on $\ell^2(\mathbb{Z})$ with positive weight sequence $\{v_n\}$ such that $I+V$ and $I+V^*$ are hypercyclic and the corresponding weights of L and V are the same except, possibly, for $v_0 \leq w_0, v_4 \leq w_4, v_{-4} \leq w_{-4}, v_{12} \leq w_{12}, v_{-20} \leq w_{-20}, v_{44} \leq w_{44}, v_{-84} \leq w_{-84}, \ldots$.

In [32], Petersson also asked whether there are other kind of dual hypercylic operators besides the one he considers. The answer is yes; the proof consists of observing that $\sigma(I+V)$ has circular symmetry but around 1 not 0. But in the case of $\ell^2(\mathbb{Z})$ we can say more. There are dual hypercylic of the form $T = I + V$, with V a bilateral weighted shift as in the theorem above and such that V is not quasinilpotent. If $\sigma(T) = \{z\colon |z-1| \leq \frac{1}{2}\}$, then by the spectral theorem $\sigma(T^n) = \{z^n\colon |z-1| \leq \frac{1}{2}\}$. For $n > 1$, these sets don't have circular symmetry with respect to a point. We can apply Corollary 2.2: all T^n are hypercyclic and T^n and T^m cannot be similar for $n \neq m$ since their spectra are different.

Problem 5.2. *Characterize those bilateral shifts V such that $I+V$ is hypercyclic.*

Feldman pointed out that when $Ve_n = v_n e_{n-1}$ with $0 < v_n \leq v_{n+1}$ for all $n \in \mathbb{Z}$ and $\lim_{n\to-\infty} v_n < 2$, then $I+V$ is hypercyclic. In this case V^* is hyponormal, and the conclusion is a consequence of his work with V. Miller and L. Miller ([16]).

Problem 5.3. *Characterize those bilateral shifts V such that $I+V$ is dual hypercyclic.*

The Volterra operator on $L^2([0,1], dx)$ is defined by $Vf(x) = \int_0^x f(t)\,dt$. Let φ be a continuous self–map of $[0,1]$. A Volterra type operator on $L^2([0,1], dx)$ is defined by $Vf(x) = \int_0^{\varphi(x)} f(t)\,dt$.

The following four results were discovered by Montes–Rodríguez, Rodríguez–Martínez and Shkarin.

Theorem 5.5. *Let H be a Hilbert space and let $\mathcal{Q} \subset \mathcal{L}(H)$ be the compact quasinilpotent operators. Then*

$$\{T = I + Q\colon Q \in \mathcal{Q} \text{ and } T \text{ is dual hypercyclic}\}$$

is a G_δ set in $\{I + Q\colon Q \in \mathcal{Q}\}$.

Proposition 5.4. *The set of w in $c_0(\mathbb{Z})$ for which $I+B_w$ is dual hypercyclic is a G_δ in $c_0(\mathbb{Z})$.*

Theorem 5.6. *Let φ be a continuous strictly increasing self–map of $[0,1]$ such that $\varphi(x) < x$ for $0 < x \leq 1$. Then $I + V_\varphi$ is hypercyclic.*

Theorem 5.7. *Let T be a continuous operator on a separable Fréchet space X such that*

$$\mathrm{Ker}^+ \, T = span(\cup_{n=1}^{\infty}(T^n(X) \cap KerT^n)$$

is dense in X. Then $I + T$ is hypercyclic.

6. Frequently hypercyclic operators

A subset A of $\mathbb{N}$ *has positive lower density* if

$$\underline{\lim}\frac{\mathrm{Card}\;(A \cap \{1, 2.., n\})}{n}) > 0.$$

An operator $T \in \mathcal{L}(B)$ is said to be *frequently hypercyclic* if there exists $x \in B$ such that, for each open V, the set $\{n \colon T^n x \in V\}$ has positive lower density. This notion was introduced recently by Bayart and Grivaux ([4]). Among their many results they have:

Theorem 6.1. *Let X be a separable Fréchet space with ρ an invariant metric, and let T be a continuous operator on X. Suppose that there exist a dense sequence (x_l) of X and a map S defined on X such that*

(1) $\sum_{k=1}^{\infty} \rho(T^k x_l, 0)$ is convergent for each l
(2) $\sum_{k=1}^{\infty} \rho(S^k x_l, 0)$ is convergent for each l
(3) $TS = I$.

Then T is frequently hypercyclic.

The first three classical examples mentioned at the beginning are frequently hypercyclic ([4]). In [22], Grosse-Erdman and Peris showed that every frequently hypercyclic operator satisfies the Hypercyclicty Criterion.

Problem 6.1. *Identify the compact subsets of $\mathbb{C}$ which are the spectra of dual hypercyclic operators. Do the same for the spectra of frequently hypercyclic operators. In both cases the underlying space must be taken into account. The most interesting case probably is for operators in a Hilbert space.*

Problem 6.2. *Characterize the frequently hypercyclic unilateral backward shifts ([4,22]). What about bilateral shifts?*

Problem 6.3. *What is the intersection of the class of dual hypercyclic operators and the class of frequently hypercyclic operators? It seems to be empty.*

Acknowledgments: I wish to thank the organizers of the III International Course of Mathematical Analysis in Andalucía, specially Professors Tomás Domínguez Benavides and Miguel Marano for such a well–run course and for inviting me.

References

1. S. I. Ansari, *Hypercyclic and cyclic vectors*, J. Funct. Anal. **128**(2) (1995), 374–383.
2. S. I. Ansari, *Existence of hypercyclic operators on topological vector spaces*, J. Funct. Anal. **148**(2) (1997), 384–390.
3. S. I. Ansari and P. S. Bourdon, *Some properties of cyclic operators*, Acta Sci. Math. (Szeged) **63**(1-2) (1997), 195–207.
4. F. Bayart and S. Grivaux, *Frequently hypercyclic operators*, Trans. Amer. Math. Soc. **358**(11) (2006), 5083-5117.
5. L. Bernal–González, *On hypercyclic operators on Banach spaces*, Proc. Amer. Math. Soc. **127**(4) (1999), 1003–1010.
6. J. P. Bès, *Invariant manifolds of hypercyclic vectors for the real scalar case*, Proc. Amer. Math. Soc. **127**(6) (1999), 1801–1804.
7. G. D. Birkhoff, *Demonstration de un theoreme elementaire sur les fonctiones entieres*, C. R. Acad. Sci. Paris, **189** (1929), 473-475.
8. J. Bonet and A. Peris, *Hypercyclic operators on non-normable Fréchet spaces*, J. Funct. Anal. **159**(2) (1998), 587–595.
9. P. S. Bourdon, *Invariant manifolds of hypercyclic vectors,* Proc. Amer. Math. Soc. **118**(3) (1993), 845–847.
10. P. S. Bourdon, *The second iterate of a map with dense orbit,* Proc. Amer. Math. Soc. **124**(5) (1996), 1577–1581.
11. P. S. Bourdon, *Orbits of hyponormal operators*, Michigan Math. J. **44**(2) (1997), 345–353.
12. P. S. Bourdon and N. S. Feldman, *Somewhere dense orbits are everywhere dense*, Indiana Univ. Math. J. **52**(3) (2003), 811–819.
13. P. S. Bourdon and J. H. Shapiro, *Cyclic Phenomena for composition operators*, Memoirs Amer. Math. Soc. (1997), Vol **125,** no. 596.
14. G. Costakis and M. Sambarino, *Topologically mixing hypercyclic operators*, Proc. Amer. Math. Soc. **132**(2) (2004), 385–389.
15. M. De la Rosa and C. J. Read, *A hypercyclic operator whose direct sum $T \oplus T$ is not hypercyclic*, preprint (2006).
16. N. S. Feldman, V. G. Miller and T. L. Miller, *Hypercyclic and supercyclic cohyponormal operators*, Acta Sci. Math. (Szeged) **68**(1-2) (2002), 303–328.
17. R. Gethner and J. H. Shapiro, *Universal vectors for operators on spaces of holomorpic functions*, Proc. Amer. Math. Soc. **100** (1987), 281–288.

18. G. Godefroy and J. H. Shapiro, *Operators with dense, invariant, cyclic vector manifolds*, J. Funct. Anal. **98** (1991), 229–269.
19. W. T. Gowers and B. Maurey, *The unconditional basic sequence problem*, J. Amer. Math. Soc. **6**(4) (1993), 851–874.
20. S. Grivaux, *Construction of operators with prescribed behaviour*, Arch. Math. (Basel) **81**(3) (2003), 291–299.
21. K. G. Grosse-Erdman, *Universal families and hypercyclic operators*, Bull. Amer. Math. Soc. (N.S.) **36**(3) (1999), 345–381.
22. K. G. Grosse–Erdmann and A. Peris, *Frequently dense orbits*, (English, French summary) C. R. Math. Acad. Sci. Paris **341**(2) (2005), 123–128.
23. G. H. Hardy and E. M. Wright, *An introduction to the theory of numbers* (fifth edition), The Clarendon Press, Oxford University Press, New York (1979).
24. D. A. Herrero, *Limits of hypercyclic and supercyclic operators*, J. Funct. Anal. **99** (1991), 179–190.
25. C. Kitai, *Invariant closed sets for linear operators*, Ph. D. Thesis, Univ. of Toronto (1982).
26. I. Halperin, C. Kitai and P. Rosenthal, *On orbits of linear operators*, J. Lond. Math. Soc., II. Ser. **31** (1985), 561–565.
27. F. León–Saavedra and V. Müller, *Rotations of hypercyclic and supercyclic operators*, Integral Equations Operator Theory **50**(3) (2004), 385–391.
28. J. Lindenstrauss and L. Tzafriri, *Classical Banach spaces. I. Sequence spaces*, Ergebnisse der Mathematik und ihrer Grenzgebiete, Vol. **92**. Springer–Verlag, Berlin–Heidelberg–New York (1977).
29. G. R. MacLane, *Sequences of Derivatives and normal families*, J. Analyse Math. **2** (1952), 72–67.
30. M. Marano and H. N. Salas, *Ansari's theorem revisited*, preprint (2007).
31. A. Montes–Rodríguez, A. Rodríguez–Martínez and S. Shkarin, *Cyclic Behavior of Volterra type operators*, preprint (2007).
32. H. Petersson, *Spaces that admit hypercyclic operators with hypercyclic adjoints*, Proc. Amer. Math. Soc. **134** (2005), 1671–1676.
33. S. Rolewicz, *On orbits of elements*, Studia Math. **32** (1969), 17–22.
34. H. N. Salas, *A hypercyclic operator whose adjoint is also hypercyclic*, Proc. Amer. Math. Soc. **112**(3) (1991), 765–770.
35. H. N. Salas, *Hypercyclic weighted shifts*, Trans. Amer. Math. Soc. **347**(3) (1995), 993–1004.
36. H. N. Salas, *Banach spaces with separable duals support dual hypercyclic operators*, Glasgow Math. J. **49**(2) (2007), 281–290.
37. J. H. Shapiro, *Notes on dynamics of linear operators*, http://www.math.msu.edu/shapiro, (2001).
38. A. L. Shields, *Weighted shift operators and analytic function theory*, Math. Survey Monographs, Vol. **12,** 49–128. Amer. Math. Soc. Providence, RI 1974. Second Printing (1979).
39. P. Walters, *An introduction to ergodic theory*, Graduate Texts in Mathematics, **79**, Springer–Verlag, New York–Berlin (1982).
40. B. Yousefi and H. Rezaei, *Hypercyclic property of weighted composition operators*, Proc. Amer. Math. Soc. **135**(10) (2007), 3263–3271.

OPERATOR SPACES: BASIC THEORY AND APPLICATIONS*

BERTRAM M. SCHREIBER

Department of Mathematics
Wayne State University
48202 Detroit, MI, USA
E-mail: bert@math.wayne.edu

1. Operator spaces and completely bounded maps

In the past twenty years, the theory of operator spaces has become a well–established area in Functional Analysis, with applications to the study of Banach and operator algebras, Harmonic Analysis, Probability Theory and Complex Analysis. In fact, some of the results now considered as embodied in this area predate the establishment of this body of knowledge. Nevertheless, these ideas seem not to be as widely known as warranted. Our purpose here is to survey some of the fundamental ideas in the field and mention several applications, in order to indicate the variety of directions in which these ideas can be applied. For proofs of the assertions in Sections 1-3, see [2,6].

The basic idea of the theory is *quantization* — one replaces functions by operators. We are interested here in looking at Banach spaces as subspaces of $B(H)$, the Banach algebra of operators on a Hilbert space H. One can then exploit the special nature of those spaces of operators to create a new category in which the maps are those bounded linear maps that preserve this special nature.

Definition 1.1. A (*concrete*) *operator space* is a closed linear subspace of $B(H)$ for some Hilbert space H.

Definition 1.2. If X is a concrete operator space in $B(H)$, let $M_{m,n}(X)$

*Presented to III Curso de Análisis Matemático en Andalucia on 7 September, 2008.

denote all $m \times n$ matrices (x_{ij}) of elements of X, normed as elements of $B(H^n, H^m)$, and set $M_n(X) = M_{n,n}(X)$, $M_{m,n}(\mathbb{C}) = M_{m,n}$, and $M_n(\mathbb{C}) = M_n$. For $x \in M_n(X)$, let $\|x\|_n$ denote its operator norm. If $x \in M_m(X)$ and $y \in M_n(X)$, let $x \oplus y$ denote the obvious element of $M_{m+n}(X)$:

$$x \oplus y = \begin{pmatrix} x & 0 \\ 0 & y \end{pmatrix} = \begin{pmatrix} (x_{ij}) & 0 \\ 0 & (y_{kl}) \end{pmatrix}.$$

Definition 1.3. The *operator–space structure* of X consists of the sequence of operator norms on $M_n(X)$, $n = 1, 2, \ldots$. They are related as in the following proposition. We refer to X *equipped with all of these norms on the spaces of matrices* as a concrete operator space.

Proposition 1.1. *Let X be a concrete operator space.*

(M1) $\|x \oplus y\|_{m+n} = \max\{\|x\|_m, \|y\|_n\}$: $x \in M_m(X), y \in M_n(X)$.
(M2) $\|\alpha x \beta\|_n \leq \|\alpha\| \, \|x\|_m \|\beta\|$, $x \in M_m(X)$, $\alpha \in M_{n,m}$, $\beta \in M_{m,n}$.

Example 1.1. Recall that $\gamma \colon H \to B(\mathbb{C}, H)$, where $\gamma(\xi)(a) = a\xi$, is an isometry and thus defines H itself as a concrete operator space, denoted H_c and called the *column space* of H.

On the other hand, if $\bar{H}$ denotes the usual conjugate space of H and $\xi \in H$, define $\rho \colon H \to \bar{H}^*$ by $\rho(\xi)(\bar{\eta}) = \langle \xi, \eta \rangle$. Again H becomes a concrete operator space, denoted H_r and called the *row space* of H.

Of course, H_c, H_r are isometric as Banach spaces, but they have *inequivalent operator space structures*, as defined below.

Definition 1.4. By an (*abstract*) *operator space* we mean a normed space X equipped with norms $\|\cdot\|_n$ on the spaces $M_n(X)$ satisfying (M1) and (M2). As we will see, there are a lot of them. We call the norms $\|\cdot\|_n$ the *operator–space matrix norms* of X and say that the normed space X has been given an *operator–space structure* if such norms on the spaces $M_n(X)$ have been assigned.

Remark 1.1.

(1) Given an (abstract) operator space X and $x \in M_n(X)$, it follows from (M1) that the natural map $x \mapsto x \oplus 0$ is an isometry of $M_n(X)$ into $M_{n+1}(X)$. So we can think of our operator space as defining a norm on $M^0_\infty(X)$ = all $\mathbb{N} \times \mathbb{N}$ matrices of elements of X with only finitely many nonzero entries.

(2) If X is an operator space, then $M_{m,n}(X)$ also has a distinguished norm as a subspace of $M_p(X)$, $p = \max\{m, n\}$. We assume this is the norm on $M_{m,n}(X)$. (M1), (M2) also hold here, when formulated appropriately.

Definition 1.5. Let X, Y be operator spaces and $\varphi\colon X \to Y$ be a bounded linear map. Then φ induces obvious maps

$$\varphi_n\colon M_n(X) \to M_n(Y), \quad \varphi_n\left((x_{ij})_{i,j=1}^n\right) = (\varphi(x_{ij}))_{i,j=1}^n.$$

It is clear from the isometric embedding of $M_n(X)$ in $M_{n+1}(X)$ and (M1) that

$$\|\varphi\| = \|\varphi_1\| \leq \|\varphi_2\| \leq \cdots.$$

We say φ is *completely bounded (c.b.)* if

$$\|\varphi\|_{\text{cb}} = \sup_n \|\varphi_n\| < \infty.$$

It is easy to check that $\|\varphi\|_{\text{cb}}$ is a norm on the space $CB(X,Y)$ of all completely bounded maps from X to Y. We call φ a *complete contraction* if $\|\varphi\|_{\text{cb}} \leq 1$, a *complete isometry* if each $\varphi_n\colon M_n(X) \to M_n(Y)$ is isometric, and a *complete isomorphism* if φ is a linear isomorphism and both φ and φ^{-1} are c.b.

The completely bounded maps are the natural morphisms in the category of operator spaces. It is easy to see that every Banach space X can be viewed as having at least one operator–space structure. Thus we have constructed a functor:

$$\{\text{Ban. spaces, bdd. lin. maps}\} \to \{\text{op. spaces, c.b. lin. maps}\}.$$

Example 1.2.

(1) If X and Y are operator spaces with $\dim X = n$ or $\dim Y = n$ and $\varphi \in B(X,Y)$, then $\varphi \in CB(X,Y)$ with

$$\|\varphi\|_{\text{cb}} \leq n\|\varphi\|.$$

(2) Let Ω be a locally compact space and X be any operator space. If $\varphi\colon X \to C_0(\Omega)$ is bounded and linear, then φ is completely bounded with $\|\varphi\|_{\text{cb}} = \|\varphi\|$.

(3) (Grothendieck) Let X, Ω be as in (2). There exists $K_G > 0$ such that if $\varphi\colon C_0(\Omega) \to X$ is linear and bounded, then

$$\varphi \in CB(C_0(\Omega), X), \quad \|\varphi\|_{\text{cb}} \leq K_G\,\|\varphi\|.$$

2. The fundamental theorem and some consequences

Theorem 2.1 (Fundamental Theorem; Z.-J. Ruan, 1987).
Let X be an abstract operator space. There is a Hilbert space H and a complete isometry Φ mapping X onto a closed subspace of $B(H)$. If X is separable, then H can be taken separable.

Applications.

(1) **Subspaces.** Any subspace Y of an operator space X has the operator–space structure it inherits from X by restriction, i.e., $M_n(Y) \subset M_n(X)$.

(2) **C^*-algebras.** By the Gelfand–Naimark Theorem, every C^*-algebra is *-isometrically isomorphic to a subalgebra of some $B(H)$.

(3) **Quotient spaces.** If Y is a closed subspace of the operator space X, then $M_n(Y)$ is a closed subspace of $M_n(X)$ and the natural map of $M_n(X)$ onto $M_n(X/Y)$ has kernel $M_n(Y)$. Thus

$$M_n(X/Y) \cong M_n(X)/M_n(Y).$$

So give $M_n(X/Y)$ the quotient norm of $M_n(X)/M_n(Y)$. These norms satisfy (M1) and (M2). Thus X/Y becomes an operator space with respect to this *quotient operator–space structure.*

(4) **Dual spaces.** If X is an operator space, then every element of the dual space X^* is c.b., as is easily seen. Moreover, it is easy to see from Example 1.2(1) of Section 1 that

$$M_n(X^*) \cong B(X, M_n) \cong CB(X, M_n); \quad f(x) = (f_{ij}(x)), \; f \in M_n(X^*).$$

Give $M_n(X^*)$ the norm it inherits from $CB(X, M_n)$. One checks easily that this gives X^* an operator–space structure, called the *dual operator–space structure* on X^*.

(5) **Mapping spaces.** More generally, let X, Y be operator spaces. Then each $\varphi = (\varphi_{ij}) \in M_n(CB(X,Y))$ determines a mapping $\varphi \colon X \to M_n(Y)$ in the obvious way. Thus

$$M_n(CB(X,Y)) \cong CB(X, M_n(Y)).$$

Now use the c.b.-norm on $CB(X, M_n(Y))$ to define $\|\varphi\|_n$ on $M_n(CB(X,Y))$. Then $CB(X,Y)$ becomes an operator space.

(6) Let G be a locally compact group, let $C^*(G)$ denote its group C^*-algebra (see Section 6) and let $B(G)$ be the Banach algebra of all linear combinations of functions of positive type (positive definite functions) on G. Then $C^*(G)^* = B(G)$, so the Fourier–Stieltjes algebra $B(G)$ has a natural operator–space structure. Hence so does the Fourier algebra $A(G)$, as a subspace of $B(G)$.

(7) **Bilinear maps.** Let X, Y, Z be operator spaces and let $Bil(X \times Y, Z)$ be the space of bounded bilinear maps $u\colon X \times Y \to Z$, i.e.,

$$\|u\| = \sup_{\|x\|\leq 1, \|y\|\leq 1} \|u(x,y)\| < \infty.$$

Given such a u, for all $x \in X$ there exists $\Phi(x) \in B(Y, Z)$:

$$\Phi(x)(y) = u(x,y); \quad \Phi \in B(X, B(Y, Z)).$$

Call u *completely bounded* if

$$\Phi(x) \in CB(Y, Z) \quad \forall\, x \in X \text{ and } \Phi \in CB(X, CB(Y, Z)).$$

This defines the space $CBil(X \times Y, Z)$ of *completely bounded bilinear maps*. In particular, taking $Z = \mathbb{C}$, if $u \in Bil(X, Y) = Bil(X \times Y, \mathbb{C})$, each $x \in X$ defines $\Phi(x) \in Y^*$ as above: $\Phi(x)(y) = u(x, y)$. Then u is c.b. if $\Phi\colon X \to Y^*$ is c.b.. This defines the space $CBil(X, Y)$.

(8) **Bimeasures.** Let $BM(\Omega_1, \Omega_2) = Bil(C_0(\Omega_1), C_0(\Omega_2))$. Elements of $BM(\Omega_1, \Omega_2)$ are called *bimeasures*. If we put together what we have said so far we get:

Theorem 2.2 (Grothendieck). *Every $u \in BM(\Omega_1, \Omega_2)$ is c.b., i.e.,*

$$BM(\Omega_1, \Omega_2) = CBil(C_0(\Omega_1), C_0(\Omega_2)).$$

3. Tensor products of operator spaces

Recall that if H_1, H_2 are Hilbert spaces, then there is a natural way to make the algebraic tensor product $H_1 \otimes H_2$ into an inner product space; namely, set

$$\langle \xi_1 \otimes \eta_1, \xi_2 \otimes \eta_2 \rangle = \langle \xi_1, \xi_2 \rangle_{H_1} \langle \eta_1, \eta_2 \rangle_{H_2}$$

and extend linearly. The completion of $H_1 \otimes H_2$ with respect to the norm induced by this inner product is called the *Hilbert–space tensor product* of H_1 and H_2, and we shall henceforth refer to this completion as $H_1 \otimes H_2$.

If $T_i \in B(H_i)$, $i = 1, 2$, then there is a unique operator $T_1 \otimes T_2 \in B(H_1 \otimes_2 H_2)$ satisfying

$$T_1 \otimes T_2(\xi \otimes \eta) = T_1\xi \otimes T_2\eta, \quad \xi \in H_1, \eta \in H_2,$$

and $\|T_1 \otimes T_2\| = \|T_1\|\, \|T_2\|$.

If X, Y are Banach spaces, consider the algebraic tensor product $X \otimes Y$. For $t \in X \otimes Y$, set

$$\|t\|_\gamma = \inf \left\{ \sum_{i=1}^n \|x_i\| \|y_i\| : t = \sum_{i=1}^n x_i \otimes y_i \right\}.$$

It is easy to check this is a norm and the completion, denoted $X \otimes_\gamma Y$, is called the *projective tensor product* of X and Y. It is the natural tensor product in the category of Banach spaces: for X, Y, Z Banach spaces,

$$B(X \otimes_\gamma Y, Z) \cong Bil(X \times Y, Z) \cong B(X, B(Y, Z)).$$

We want to *quantize* this notion on operator spaces.

Proposition 3.1. *Let X, Y be vector spaces.*

(1) $M_p(X) \otimes M_q(Y) \cong M_{pq}(X \otimes Y)$ (Kronecker product).
(2) If $t \in M_n(X \otimes Y)$, then there exist integers p and q, $x \in M_p(X)$, $y \in M_q(Y)$, $\alpha \in M_{n,pq}$ and $\beta \in M_{pq,n}$ such that

$$t = \alpha(x \otimes y)\beta.$$

Definition 3.1. Let X and Y be operator spaces. For $t \in M_n(X \otimes Y)$, set

$$\|t\|_\Gamma = \inf\{\|\alpha\|\,\|x\|\,\|y\|\,\|\beta\| : t = \alpha(x \otimes y)\beta \text{ as in Proposition 3.1}\}.$$

As asserted in the following theorem, this formula defines a norm on $M_\infty^0(X \otimes Y)$. The completion $X \widehat{\otimes} Y$ with respect to this norm is called the *operator–space projective tensor product* of X and Y.

Theorem 3.1. *For any operator spaces X and Y, $X \widehat{\otimes} Y$ is an operator space with $\|x \otimes y\|_\Gamma \leq \|x\|_m \|y\|_n$ for all $x \in M_m(X), y \in M_n(Y)$.*

In fact, $\widehat{\otimes}$ is commutative and associative and is the natural tensor product in the category of operator spaces:

Proposition 3.2. *If X, Y, and Z are operator spaces, there are complete isometries*

$$CB(X \widehat{\otimes} Y, Z) \cong CBil(X \times Y, Z) \cong CB(X, CB(Y, Z)).$$

Corollary 3.1. *If X and Y are operator spaces then we have as a complete isometry*

$$\left(X \widehat{\otimes} Y\right)^* = CB(X, Y^*),$$

where each $u \in \left(X \widehat{\otimes} Y\right)^*$ corresponds to the canonical map $\varphi_u \colon X \to Y^*$ given by

$$\varphi_u(x)(y) = u(x \otimes y).$$

There is another notion of complete boundedness for multilinear maps which plays an important role in operator space theory and its applications. It is defined by mimicking the multiplication of matrices.

Definition 3.2. Let $\varphi \in CBil(X \times Y, Z)$ and $\bar{\varphi}\colon X \otimes Y \to Z$ be its linear extension. Then, as we have seen, there correspond maps

$$\varphi_{n,n}\colon M_n(X) \times M_n(Y) \to M_{n^2}(Z)$$

given by

$$\varphi_{n,n}(x,y)_{(i,k),(j,l)} = \bar{\varphi}_{n^2}(x \otimes y)_{(i,k),(j,l)} = \varphi(x_{ij}, y_{kl}).$$

Define $\varphi_{(n)}\colon M_n(X) \times M_n(Y) \to M_n(Z)$ by

$$\left[\varphi_{(n)}(x,y)\right]_{i,j} = \sum_{k=1}^{n} \varphi(x_{ik}, y_{k,j}), \quad 1 \leq i,j \leq n.$$

We call φ *matrix completely bounded* (m.b.) if

$$\|\varphi\|_{\text{mb}} = \sup_n \|\varphi_{(n)}\| < \infty.$$

If $\|\varphi\|_{\text{mb}} \leq 1$ we say φ is a *matrix complete contraction.*

Let $MBil(X \times Y, Z)$ be the linear space of all matrix c.b. maps from $X \times Y$ to Z with the mb-norm. If $\varphi \in CBil(X \times Y, Z)$, then one can check that $\|\varphi\|_{\text{cb}} \leq \|\varphi\|_{\text{mb}}$. Thus

$$MBil(X \times Y, Z) \subset CBil(X \times Y, Z).$$

Let us now linearize the m.b. maps, just as we did for the c.b. maps to create the operator–space projective tensor product. To do this, replace the Kronecker product $x \otimes y$ by the "matrix inner product"— replace matrix multiplication by tensor product, as follows.

Definition 3.3. If $x \in M_{m,p}(X), y \in M_{p,n}(Y)$, define

$$x \odot y \in M_{m,n}(X \otimes Y)$$

by

$$(x \odot y)_{ij} = \sum_{k=1}^{p} x_{ik} \otimes y_{kj}.$$

Thus if $\varphi\colon X \times Y \to Z$ is bilinear, then

$$\varphi_{(n)}(x,y) = \bar{\varphi}(x \odot y).$$

Proposition 3.3. *Let X, Y be vector spaces. If $t \in M_n(X \otimes Y)$, then there exist $p \geq 1$, $x \in M_{n,p}(X)$ and $y \in M_{p,n}(Y)$ such that $t = x \odot y$.*

Definition 3.4. Let X and Y be operator spaces. For $t \in M_n(X \otimes Y)$, set

$$\|t\|_h = \inf\{\|x\|\|y\| : t = x \odot y,\ p, x, y \text{ as in Proposition 3.3}\}.$$

By Proposition 3.3, the set on the right is nonempty. We now assert that this defines an operator–space structure on $X \otimes Y$, whose completion $X \otimes_h Y$ is called the *Haagerup tensor product* of X and Y.

Theorem 3.2. *Let X and Y be operator spaces. Then $\|\cdot\|_h$ defines an operator–space structure on $X \otimes Y$ such that for any t in $M_n(X \otimes Y)$, $\|t\|_h \leq \|t\|_\Gamma$.*

Corollary 3.2. $MBil(X \times Y, Z) \cong CB(X \otimes_h Y, Z)$.

Remark 3.1. The Haagerup tensor product is associative but not commutative.

Theorem 3.3 (U. Haagerup). *Let A and B be (unital) C^*-algebras and suppose that $u \in (A \otimes_h B)^*$. Then there exist *-representations*

$$\pi \colon A \to B(H_\pi),\ \sigma \colon B \to B(H_\sigma),\ \xi \in H_\sigma,\ \eta \in H_\pi,$$

and $T \colon H_\sigma \to H_\pi$ such that

$$u(a \otimes b) = \langle \pi(a)T\sigma(b)\xi, \eta\rangle = \langle T\sigma(b)\xi, \pi(a^*)\eta\rangle \tag{1}$$

and $\|u\| = \|T\|\|\xi\|\|\eta\|$.

Conversely, if $\varphi \colon A \otimes_h B \to \mathbb{C}$ is given as in the right–hand side of (1), *then $\varphi \in (A \otimes_h B)^*$ with $\|\varphi\| \leq \|T\|\|\xi\|\|\eta\|$.*

Remark 3.2.

(1) Theorem 3.3 implies an improvement of Theorem 2.2:

Theorem 3.4 (Grothendieck). *Let Ω_1, Ω_2 be locally compact spaces. Then every bounded bilinear form on $C_0(\Omega_1) \times C_0(\Omega_2)$ is matrix completely bounded.*

(2) Let $X_1, \ldots, X_n$ be operator spaces. It is simple to extend Definition 3.2 to define matrix completely bounded multilinear maps

$$\varphi \colon X_1 \times \cdots \times X_n \to Y.$$

(3) Using the notion of a unitary dilation and a nontrivial generalization of Theorem 3.3, we obtain the following theorem:

Theorem 3.5 (E. Christensen, A. Sinclair). *Let $A_1, \dots, A_n$ be C^*-algebras, K be a Hilbert space, and*

$$\varphi\colon A_1 \times \cdots \times A_n \to B(K)$$

be multilinear. Then φ is m.b. if and only if there is a Hilbert space H, $*$-representations $\pi_i\colon A_i \to B(H)$, $1 \le i \le n$, $S \in B(K,H)$ and $T \in B(H,K)$ such that $\|\varphi\|_{\mathrm{mb}} = \|S\|\|T\|$ and

$$\varphi(a_1, \dots, a_n) = T\pi_1(a_1)\cdots\pi_n(a_n)S.$$

(4) It can be shown that $\otimes_h$ is *injective* in the category of operator spaces: if $X_1 \subset X_2$, $Y_1 \subset Y_2$ then $X_1 \otimes_h Y_1 \subset X_2 \otimes_h Y_2$ completely isometrically. This and a Hahn–Banach Theorem argument can be used to extend all of the above results from C^*-algebras to all operator spaces.

4. Applications to operator algebra theory

4.1. *An application of the operator–space projective tensor product*

Recall that a *von Neumann (v.N.) algebra* M on a Hilbert space H is a C^*-subalgebra of $B(H)$ that is equal to its *double commutant*:

$$M' = \{t \in B(H)\colon tm = mt, m \in M\},$$

$M'' = (M')'$ and we require that $M'' = M$.

By a well–known theorem of Sakai, a C^*-algebra A can be realized as a v.N. algebra in some $B(H)$ if and only if A is isometrically the dual space of some Banach space. In particular if A is embedded in some $B(H)$, then $A'' = A^{**}$, so A^{**} is a v.N. algebra.

The predual A_* of a v.N. algebra is unique. For example, it is well–known that $B(H)_* = \mathcal{T}(H)$ (trace–class operators).

Let M be a v.N. algebra. Then M_* has a natural operator–space structure as a subspace of M^*: $M_* \subset (M_*)^{**} = M^*$. And, in fact, the dual–space operator–space structure induced by M_* on M is the original one.

Definition 4.1. Let H, K be Hilbert spaces and $M \subset B(H)$ and $N \subset B(K)$ be v.N. algebras with preduals M_* and N_*, respectively. Let $M\overline{\otimes}N$ be the v.N. algebra closure of

$$M \otimes N = \{m \otimes n\colon m \in M, n \in N\} \subset B(H \otimes K).$$

So $M\overline{\otimes}N = (M \otimes N)'' = \overline{M \otimes N}^{\text{wk}*}$. It is known that this is (up to equivalence) independent of the realizations and that

$$M_* \otimes N_* \text{ is dense in } (M\overline{\otimes}N)_*.$$

Theorem 4.1. *There is a complete isomorphism from $M\overline{\otimes}N$ onto $CB(M_*, N)$.*

Corollary 4.1. *As a complete isomorphism,*

$$(M\overline{\otimes}N)_* \cong M_*\widehat{\otimes}N_*.$$

Proof. By Theorem 4.1 and Corollary 3.1,

$$M\overline{\otimes}N = CB(M_*, N) = CB(M_*, (N_*)^*) \cong (M_*\widehat{\otimes}N_*)^*.$$

That is, $M_*\widehat{\otimes}N_*$ is the (unique) predual of $M\overline{\otimes}N$. □

4.2. *An application to nonselfadjoint operator algebras*

Definition 4.2. An *injective operator space* is the range of an idempotent complete contraction P on some $B(H)$. Such a space is a *ternary ring of operators* (TRO), i.e., a closed subspace Z of a C^*-algebra such that $ZZ^*Z \subset Z$. A *ternary morphism* T between TRO's is a map satisfying

$$T(xy^*z) = T(x)T(y)^*T(z).$$

Theorem 4.2 (M. A. Youngson). *The range of an idempotent complete contraction P on a TRO Z is (ternary isomorphic to) a TRO with new triple product*

$$(x, y, z) \mapsto P(xy^*z), \;\; x, y, z \in P(Z).$$

Definition 4.3. For any subspace X of $B(H)$, there is a subspace $I(X) \supset X$ which is injective and contains no smaller injective space containing X. We call $I(X)$ an *injective envelope* of X. $I(X)$ is unique up to ternary isomorphisms.

Theorem 4.3 (M. Kaneda, V. Paulsen). *A Banach space X has the structure of an operator algebra if and only if it is an operator space such that when it is represented as a concrete operator space there exists $u \in I(X)$ such that $Xu^*X \subset X$. To define the multiplication, make $\|u\| = 1$ and set $x \cdot y = xu^*y$.*

Proof. (D. Blecher) If X satisfies the given conditions, let

$$\theta(x) = \begin{pmatrix} xu^* & x(1-u^*u)^{1/2} \\ 0 & 0 \end{pmatrix}.$$

Then θ is a complete isometric algebra homomorphism.

If A is a subalgebra of $B(H)$ and $I(A) = P(B(H))$ is an injective envelope for A, set $u = P(1)$. By Theorem 4.2,

$$xy = P(x1^*y) = P(xP(1)^*y) = xu^*y. \qquad \square$$

5. Applications to harmonic analysis

5.1. *An application to Fourier algebras*

Let G be a locally compact group, with right Haar (translation–invariant) measure dx. Let λ denote the right regular (unitary) representation of G on $L^2(G, dx)$:

$$(\lambda(x)f)(y) = f(yx) \text{ a.e.}, \quad f \in L^2(G),\ x, y \in G.$$

Denote the linear space of all "matrix coefficients" of λ, i.e., all functions of the form

$$F(x) = \langle \lambda(x)f, g\rangle = f * g^*, \quad x \in G, \tag{2}$$

by $A(G)$, and norm $A(G)$ by

$$\|F\| = \inf\{\|f\|_2\|g\|_2 \colon F \text{ is represented as in (2)}\}.$$

It can be shown that $A(G)$, under pointwise operations on G and this norm, becomes a commutative Banach algebra whose Gelfand space is G via point evaluations.

Let $VN(G)$ be the v.N. algebra in $B(L^2(G))$ generated by $\lambda(G)$. Then

$$A(G)^* = VN(G).$$

Indeed, if $\Phi \in A(G)^*$ one can find $T \in VN(G)$ with $\|T\| = \|\Phi\|$ such that

$$\Phi(F) = \langle Tf, g\rangle, \quad F \text{ as in (2)}, \tag{3}$$

and every $T \in VN(G)$ defines such a Φ via (3). $VN(G)$ is called the *reduced group von Neumann algebra* of G.

Definition 5.1. A *quantized Banach algebra* is a Banach algebra which is also an operator space such that the multiplication operation $m\colon A \times A \to A$ is completely bounded.

It can be shown (but it is not obvious) that $A(G)$ is a quantized Banach algebra.

Now, if G, H are locally compact groups, it is clear that

$$A(G) \otimes_\gamma A(H) \subset A(G \times H).$$

It was shown long ago by V. Losert that this containment can be proper. However:

Theorem 5.1 (E. Effros, Z.-J. Ruan). *Let G, H be locally compact groups. As a complete isomorphism of quantized Banach algebras,*

$$A(G \times H) \cong A(G)\widehat{\otimes}A(H).$$

Proof. It is easy to check that

$$VN(G)\overline{\otimes}VN(H) \cong VN(G \times H).$$

So, by Corollary 4.1,

$$A(G \times H) \cong VN(G \times H)_* \cong VN(G)_*\widehat{\otimes}VN(H)_* \cong A(G)\widehat{\otimes}A(H).$$

It is easy to see that the multiplication on $A(G) \otimes A(H)$ agrees with that of $A(G \times H)$. As observed above, $A(G) \otimes A(H)$ is dense in $A(G \times H)$. So, by completing, we finish the proof. □

The algebras $A(G)$ and some of their relatives are currently of great research interest.

5.2. *An application of the Haagerup tensor product*

Definition 5.2. Let $\Omega_1, \ldots, \Omega_n$ be locally compact spaces. Set

$$CB(\Omega_1, \ldots, \Omega_n) = (C_0(\Omega_1) \otimes_h \cdots \otimes_h C_0(\Omega_n))^*.$$

Equivalently, $CB(\Omega_1, \ldots, \Omega_n)$ is the space of all m.b. n-linear forms on $C_0(\Omega_1) \times \cdots \times C_0(\Omega_n)$. Recall that when $n = 2$, Grothendieck's theorem says this includes all bounded bilinear forms (bimeasures). In $C_0(\Omega_i)^{**} = M(\Omega_i)^*$, let $\mathcal{L}^\infty(\Omega_i)$ denote all bounded, Borel–measurable functions. Also, we may identify $f_1 \otimes \cdots \otimes f_n$ with the function

$$(f_1 \otimes \cdots \otimes f_n)(x_1, \ldots, x_n) = f_1(x_1)f_2(x_2)\cdots f_n(x_n).$$

Note that every $u \in CB(\Omega_1, \dots \Omega_n)$ extends completely isometrically to $(\mathcal{L}^\infty(\Omega_1) \otimes_h \cdots \otimes_h \mathcal{L}^\infty(\Omega_n))^*$ in a canonical way.

Definition 5.3. For $\Omega_1, \dots, \Omega_n$ as above, and $\mu \in M(\Omega_1 \times \cdots \times \Omega_n)$, let

$$u_\mu(f_1 \otimes \cdots \otimes f_n) = \int_{\Omega_1 \times \cdots \times \Omega_n} f_1(x_1) \cdots f_n(x_n)\, d\mu(x_1, \dots, x_n).$$

Then one can show that $u_\mu \in CB(\Omega_1, \dots, \Omega_n)$.

Theorem 5.2 (G. Zhao, B. M. Schreiber). *Let $G_1, \dots, G_n$ be locally compact groups. There is a natural convolution multiplication and adjoint operation defined on $CB(G_1, \dots, G_n)$ making it into a unital Banach *-algebra which extends the *-algebra structure of $M(G_1 \times \cdots \times G_n)$ via Definition 5.3.*

These algebras behave in some ways like measure algebras and in some ways differently. Some things are known about the harmonic analysis of these convolution algebras, but there are many open questions.

6. Applications to Probability Theory

Let $(\Omega, \mathcal{A}, P)$ be a probability space and H be a Hilbert space. Consider a stochastic process (random field)

$$\mathcal{X} = \{X_t \colon t \in G\}$$

on a locally compact group G with values in $\mathcal{H} = L^2(\Omega, P; H)$. Assume that $t \mapsto X_t$ is continuous and $\mathcal{X}$ spans $\mathcal{H}$.

$\mathcal{X}$ is called (*weakly right*) *stationary* if

$$E\langle X_{sg}, X_{tg}\rangle = E\langle X_s, X_t\rangle, \quad s, t, g \in G. \tag{4}$$

If we look at the *covariance function*

$$\kappa(s,t) = E\langle X_s, X_t\rangle = \int_\Omega \langle X_s, X_t\rangle\, dP, \tag{5}$$

then stationarity says

$$\kappa(s,t) = E\langle X_{st^{-1}}, X_e\rangle = \tilde{\kappa}(st^{-1}). \tag{6}$$

Equations (5) and (6) imply that $\tilde{\kappa}$ is of positive type on G, so κ has spectral properties, which are used heavily in applications.

A great deal is known about stationary processes. For instance, if G is abelian with character group $\widehat{G}$, then, by Bochner's Theorem, there is an *associated spectral measure* μ on $\widehat{G}$ such that

$$\tilde{\kappa}(t) = \hat{\mu}(t) = \int_{\widehat{G}} \gamma(t)\, d\mu(\gamma), \quad t \in G,$$

so

$$\kappa(s,t) = \hat{\mu}(st^{-1}) = \int_{\widehat{G}} \gamma(s)\gamma(t^{-1})\, d\mu(\gamma).$$

We want to consider *nonstationary* processes which retain some connection to Harmonic Analysis, i.e, to the spectral representation theory.

Let $C^*(G)$ be the *group C^*-algebra* of G. That is, there is a $*$-representation π_0 of the convolution algebra $L^1(G)$ on a Hilbert space H_0 such that

$$\|\pi(f)\| \le \|\pi_0(f)\|$$

for every such representation π. If we set

$$\|f\|_0 = \|\pi_0(f)\|, \quad f \in L^1(G),$$

then $\|f\|_0$ is a C^*-algebra norm, and $C^*(G)$ is the completion of $L^1(G)$ with respect to this norm. We call π_0 the *universal representation* of $L^1(G)$.

If G is abelian, we can take $C^*(G) = C_0(\widehat{G})$ and $\pi_0(f) = \widehat{f}$ (Fourier transform), $f \in L^1(G)$.

Let $W^*(G)$ be the v.N. algebra generated by $C^*(G)$ in $B(H_0)$. It is called the *group von Neumann algebra* of G. In fact, $\pi_0(x) \in W^*(G)$, $x \in G$.

If u is a bounded bilinear form on $C^*(G) \times C^*(G)$, canonically lift u from $C^*(G) \times C^*(G)$ to $W^*(G) \times W^*(G)$. Define

$$\hat{u}(s,t) = u(\pi_0(s), \pi_0(t)), \quad s,t \in G.$$

Definition 6.1. We say that $\mathcal{X} = \{X_t \colon t \in G\}$ is *V-bounded* if for some $C > 0$,

$$\left\| \int_G f(t) X_t\, dt \right\|_{\mathcal{H}} \le C\|\pi_0(f)\|, \quad f \in L^1(G).$$

The process $\mathcal{X}$ is called (*weakly*) *harmonizable* if there is a bounded bilinear form u on $C^*(G) \times C^*(G)$ such that

$$\kappa(s,t) = \hat{u}(s,t^{-1}), \quad s,t \in G.$$

Theorem 6.1 (M. M. Rao, H. Niemi, K. Ylinen). *The following are equivalent:*

(1) $\mathcal{X}$ *is V-bounded.*
(2) $\mathcal{X}$ *is harmonizable.*

If G is abelian, then (1) *and* (2) *are equivalent to:*

(3) $\mathcal{X}$ *is truly "harmonizable" in the sense that there is a vector measure* Z *on* $\widehat{G}$ *with values in* $\mathcal{H}$ *such that*

$$X_t = \int_{\widehat{G}} \gamma(t)\, dZ(\gamma), \quad t \in G.$$

If the bilinear form u in (2) *is matrix completely bounded (which is always true when G is abelian, by Theorem 3.4), then* (1) *and* (2) *are equivalent to:*

(4) There exists a Hilbert space $\mathcal{K} \supset \mathcal{H}$ *and a stationary process* $\mathcal{Y} = \{Y_t\colon t \in G\}$ *with values in* $\mathcal{K}$ *such that* $X_t = P_{\mathcal{H}} Y_t$, $t \in G$, *where* $P_{\mathcal{H}}$ *is the canonical projection onto* $\mathcal{H}$.

A more quantized notion of harmonizability can be obtained as follows.

Definition 6.2. Let $\mathcal{T}(H)$ denote the ideal of trace–class operators on H. For $x, y \in \mathcal{H}$, let the *operator inner product* $[x, y]$ be defined as a vector integral of rank–one operators by

$$[x, y] = \int_{\Omega} x(\omega) \otimes y(\omega)\, dP(\omega),$$

i.e.,

$$\langle [x, y]\xi, \eta\rangle = \int_{\Omega} \langle (x \otimes y)\,\xi, \eta\rangle\, dP = \int_{\Omega} \langle \xi, y\rangle \langle x, \eta\rangle\, dP, \quad \xi, \eta \in H.$$

The following assertions are easy to check.

Proposition 6.1. *Let* $x, y, z \in \mathcal{H}$ *and* $\lambda \in \mathbb{C}$.

(1) $[x, y] \in \mathcal{T}(H)$.
(2) $[x, x] \geq 0$ *(positive operator) and* $[x, x] = 0$ *if and only if* $x = 0$.
(3) $[x + y, z] = [x, z] + [y, z]$ *and* $[\lambda x, y] = \lambda [x, y]$.
(4) $[y, x] = [x, y]^*$.
(5) $\operatorname{tr}[x, y] = \langle x, y\rangle_{\mathcal{H}}$ *and* $\|[x, x]\| = \|x\|^2_{\mathcal{H}}$.

Definition 6.3. For $\mathcal{X}$ as above, the *operator covariance function* of $\mathcal{X}$ is

$$K(x, t) = [X_s, X_t]$$

Definition 6.4. We call $\mathcal{X}$ *operator stationary* if K satisfies the analogue of Equation (4) and (*weakly*) *operator harmonizable* if there is a bounded bilinear form $U\colon C^*(G) \times C^*(G) \to \mathcal{T}(H)$ such that

$$K(s,t) = \hat{U}(s,t^{-1}) = U(\pi_0(s), \pi_0(t)^*)$$

(Recall $\mathcal{T}(H)^* = \mathcal{B}(H)$, so

$$U = U^{**}\colon C^*(G) \times C^*(G) \to \mathcal{B}(H)^* \supset \mathcal{T}(H)).$$

One can define operator V-boundedness and obtain an analogue of Theorem 6.1 in this context, using ideas from operator–space theory. Other classes of stochastic processes are waiting to be investigated via operator–space techniques.

References

1. D. P. Blecher and C. Le Merdy, *Operator algebras and their modules–An operator space approach*, Clarendon Press, Oxford (2004).
2. E. G. Effros and Z.-J. Ruan, *Operator spaces*, Clarendon Press, Oxford (2000).
3. J. E. Gilbert, T. Ito and B. M. Schreiber, *Bimeasure algebras on locally compact groups*, J. Functional Anal. **64** (1985), 134–162.
4. C. C. Graham and B. M. Schreiber, *Bimeasure algebras on LCA groups*, Pacific J. Math. **115** (1984), 91–127.
5. Y. Kakihara, *Multidimensional Second Order Stochastic Processes*, Ser. on Multivariate Anal. vol. 2, World Scientific, Singapore (1996).
6. G. Pisier, *Introduction to Operator Space Theory*, Lond. Math. Soc. Lect. Notes Ser. **294**, Cambridge Univ. Press, Cambridge (2003).
7. V. Runde, *Applications of operator spaces to abstract harmonic analysis*, Expo. Math. **22** (2004), 317–363.
8. B. M. Schreiber, *Asymptotically stationary and related processes*, Lect. Notes in Pure and Appl. Math. vol. 238, Marcel Dekker (2004), 363–397.
9. G. Zhao, *Completely bounded multilinear forms on $C(X)$-spaces*, Proc. Roy. Irish Acad. **96A** (1996), 111–122.
10. G. Zhao and B. M. Schreiber, *Algebras of multilinear forms on groups*, Contemporary Math. **189** (1995), 497–511.

MATHEMATICS AND MARKETS: EXISTENCE AND EFFICIENCY OF COMPETITIVE EQUILIBRIUM

ANTONIO VILLAR

Department of Economics
Universidad Pablo de Olavide
41013 Seville, Spain
E-mail: avillar@upo.es

1. Introduction

Most societies organize their economic activity through the functioning of markets. Myriads of individual economic agents make decisions according to their private interests, whose interaction results in an allocation of resources. The production and exchange of commodities is at the centre of the picture: consumers demand commodities and supply labour services, firms produce commodities according to their technological knowledge, and commodities flow among agents by means of an exchange process which is realized through markets and prices.

General equilibrium models try to capture the logic of this complex network of interactions viewing the economic system as a whole, that is, taking all the simultaneous interdependencies established among economic agents into account (as opposed to partial equilibrium models, that typically concentrate on the analysis of specific markets or decision units). The first concern of general equilibrium theory is the analysis of conditions ensuring that all the actions taken independently by economic agents are simultaneously feasible. An *equilibrium* is a situation in which all the agents are able to realize their plans simultaneously; in other words, agents do not find it beneficial to change their actions.

Note that nothing ensures that the feasibility of the collective action corresponds to a socially desirable state of affairs. That is why the analysis of the social desirability of equilibrium outcomes comes next in the agenda. Suppose that the economy is arranged in such a way that all agents are simultaneously realizing their plans. Can the economy do better? If this were the case, there would be scope for the intervention of some authority (the Government, say), because changing the spontaneous allocation of resources would result in a better state. The key questions are, of course, what "better" means, and whether such an authority will be able to improve the social situation.

There are many ways of ranking the outcomes of an economy, but there is a simple principle which seems difficult to object: no resource allocation can be considered satisfactory if it were possible to improve the situation of all the members of the society with the available resources. This is the Pareto principle, which is to be understood as a minimal test of economic *efficiency.* Note that there may well be allocations passing this test which still can be deemed socially undesirable. To be clear: we are not saying that the Pareto principle ensures good outcomes; what we are saying is that one should be worried about those outcomes which do not pass such a simple test.

We present here a simplified general equilibrium model of a (pure exchange) competitive economy. A competitive economy is one in which individual agents have no market power. In particular, each individual agent takes market prices as external parameters over which she has no influence (price–taking behaviour). This setting describes a world made of many agents each of which is very small with respect to the global market. We shall concentrate here on the case of "pure exchange economies" to make things simpler (even though the argument extends immediately to the case of production economies). We present here the basic standard results: (i) A competitive equilibrium exists, under fairly reasonable assumptions; (ii) Competitive equilibria yield efficient allocations (First Welfare Theorem); and (iii) Any efficient allocation can be realized as competitive equilibria (Second Welfare Theorem).

Those results may be regarded as the *Invisible Hand Theorem*, a summary of the most relevant features of competitive markets: competitive equilibria constitute a non-empty subset of the set of efficient allocations. The idea that markets are adequate institutions for the efficient allocation of resources in a decentralized way is a very old one. It was first conceptually formulated by Adam Smith in 1776 ([11]). Lon Walras (1784) gave a formal statement of this problem ([13]). It took more than fifty years to find a proper answer, helped by the development of some mathematical tools, such as convex analysis, non-linear programming and fixpoint theory.

2. Social equilibrium

Let us present first a very abstract setting in which those questions can be formally framed. This reference model will help us to understand the nature of the requirements of general equilibrium models for competitive markets.

We consider a society made of h agents. An agent is a decision unit, that is, an individual or an institution that makes decisions over some feasible set, according to some individual goal. We assume that agents are "rational", meaning that they make the best possible choices according to their own goals. Note that this type of behaviour contains the ingredients of a standard optimization problem: maximize some objective function (that summarizes the agent's goals) within some feasible set (that describes the environmental restrictions).

An *equilibrium* in this context is a situation (a collection of decisions) in which all the actions chosen by individual agents are compatible. That is, they are individual best attainable choices that are collectively feasible. The difficulty in ensuring the existence of such an array of actions comes from the fact that agents' choices are interdependent. On the one hand, because we require collective feasibility. On the other hand, because an agent's feasible set may be conditioned by the actions of other agents.

Let us formalize these ideas and point out the implications of the way of modelling.

A *society* is a collection of h agents. Agent $i = 1, 2, \ldots, h$ is characterized by three different elements: (i) Her *choice* set A_i (the universe of alternatives in which she has to choose); (ii) her *objective function*, v_i, that embodies her choice criterion; and (iii) the *restrictions* that the agent faces, $\gamma_i(\cdot)$ (a subset of the choice set that may change with the actions of other agents).

We take $\mathbb{R}^l$ as the reference space, that is $A_i \subset \mathbb{R}^l$, so that making a choice amounts to selecting a vector in $\mathbb{R}^l$. The agent's objective function $v_i \colon A_i \to \mathbb{R}$ is a real valued function. This implies, in particular, that all choices are ordered because we associate a real number to any option in A_i (hence, v_i gives a full description of how this agent ranks the different alternatives)*. The mapping $\gamma_i \colon \prod_{k=1}^{h} A_k \to A_i$ that describes the *restrictions* faced by the agent, is a correspondence (a set–valued mapping) that depends on the actions of other agents.

*A more general approach would be to allow people's objective functions to depend not only on their own choices but also on the choices made by others. This implies that v_i is a real valued function defined on the Cartesian product of all agents' choice sets, $v_i \colon \prod_{k=1}^{h} A_k \to \mathbb{R}$. No special difficulty derives from this more general model concerning the existence of equilibrium

A *society* (also called an *abstract economy*) can thus be summarized by a tuple $(A_i, v_i, \gamma_i)_{i=1}^h$. A point in the choice set of agent i is denoted by $a_i \in A_i$, whereas $a = (a_1, \ldots, a_h)$ denotes a point in $\prod_{k=1}^h A_k$ (that is, an array of actions, one for each agent). For each $a \in \prod_{k=1}^h A_k$, the set $\gamma_i(a) \subset A_i$ defines agent i's feasible set. Note that the very notion of feasible set implies that γ_i is actually independent on its i-th coordinate vector.

We say that a point $a \in \prod_{k=1}^h A_k$ is *collectively feasible* if $a \in \prod_{k=1}^h \gamma_k(a)$. That is, a collection of individual alternatives is jointly feasible when they are consistent with the restrictions they impose.

The rational behaviour of agent i can be described by the following program:

$$\left.\begin{array}{l} \max_{a_i} v_i(a_i) \\ \text{s. t.: } a_i \in \gamma_i(a) \end{array}\right\} \qquad [\text{P}]$$

We denote by $\mu_i \colon \prod_{k=1}^h A_k \to A_i$ the (typically set–valued) mapping that associates agent i's best response to environment a. That is, given the actions chosen by all agents other than i, this agent's rational behaviour results in the selection of some option in the set:

$$\mu_i(a) = \{a_i \in \gamma_i(a) \colon v_i(a) \geq v_i(\bar{a}_i), \forall \bar{a}_i \in \gamma_i(a)\}.$$

That is, agent i maximizes her objective function on her feasible set, which is determined by others' decisions.

We can now introduce the notion of social equilibrium. It corresponds to a collection of decisions that is collectively feasible and such that every agent is maximizing her objective function. Formally:

Definition 2.1. A *social equilibrium* for a society $[A_i, v_i, \gamma_i]_{i=1}^h$ is a point $a^* \in \prod_{k=1}^h A_k$ such that:

(i) $a^* \in \prod_{k=1}^h \gamma_k(a^*)$;
(ii) $v_i(a_i^*) \geq v_i(a_i)$, $\forall a_i \in \gamma_i(a^*), i = 1, 2, \ldots, h$.

Ensuring the existence of a social equilibrium in this general context is a very demanding quest. It amounts to knowing that we can solve h simultaneous and interdependent optimization programs [P] with virtually no data. Proving the existence of equilibrium calls, therefore, for powerful tools. The most natural one in this context is the recourse to a fixed point argument. This can be better understood as follows.

First note that a point $a^* \in \prod_{k=1}^h A_k$ is a social equilibrium if and only if $a_i^* \in \mu_i(a^*)$ for all i. Now let $\mu\colon \prod_{k=1}^h A_k \to \prod_{k=1}^h A_k$ be a mapping defined by $\mu(a) = \prod_{i=1}^h \mu_i(a)$. Then, a^* is a social equilibrium if and only if $a^* \in \mu(a^*)$, that is, if and only if a^* is a fixed point of the correspondence μ.

Kakutani's fixed point theorem is the key mathematical tool to solve our problem. It establishes that any upper–hemicontinuous correspondence that applies a compact convex set over itself has a fixed point (see [3] for some variants and extensions)†.

Consider now the following set of assumptions that will allow to apply this theorem and ensure the existence of social equilibria.

- A.1.- $A_i \subset \mathbb{R}^l$ is non-empty, compact and convex.
- A.2.- $v_i\colon \prod_{k=1}^h A_k \to \mathbb{R}$ is a continuous, quasiconcave function‡.
- A.3.- $\gamma_i\colon \prod_{k=1}^h A_k \to A_i$ is a continuous correspondence with non-empty, closed and convex values§.

The following theorem provides the basic result for the existence of equilibrium:

Theorem 2.1. *Let $[A_i, v_i, \gamma_i]$ be a society and suppose that assumptions (A.1) to (A.3) hold. Then a social equilibrium exists.*

Proof. First note that, under the conditions established (a continuous objective function and a compact non-empty feasible set), Weierstrass' Theorem ensures that program [P] has a solution. That is, $\mu_i(a) \neq \emptyset$ for all a. Moreover, the quasiconcavity and continuity of the objective function imply that $\mu_i(a)$ is convex and closed.

Next we show that, for each $i = 1, 2, \ldots, h$, the correspondence μ_i is upper hemicontinuous (this is just an application of the Maximum Theorem). As A_i is compact, it suffices to show that for all sequences $\{a^n\} \subset \prod_{k=1}^h A_k$,

---

†A set mapping $\alpha\colon D \subset \mathbb{R}^n \to Y \subset \mathbb{R}^k$ is upper–hemicontinuous if, for all sequences $\{x^n\} \subset D, \{y^n\} \subset Y$ converging to x^0 and y^0, respectively, and such that $y^n \in \alpha(x^n)$ for all n, it follows that $y^0 \in \alpha(x^0)$.

‡A real valued function $f\colon C \subset \mathbb{R}^n \to \mathbb{R}$ is called quasiconcave if $f(x) > f(x')$ implies that $f[\lambda x + (1-\lambda)x')] > f(x')$ for all $\lambda \in (0,1)$. This property is a substantial generalization of the concavity notion and implies that the upper level sets are convex. Quasiconcavity is a postulate on the *liking of variety*: intermediate combinations of choices tend to be more appreciated.

§A correspondence is *continuous* when it is both upper and lower hemicontinuous (where lower hemicontinuity requires that each point $y^0 \in \alpha(x^0)$ be approachable by a sequence of points $\{x^n\} \subset D$, $\{y^n\} \subset Y$, with $y^n \in \alpha(x^n)$).

$\{a_i^n\} \subset A_i$, converging to a^0 and a_i^0, respectively, and such that $a_i^n \in \mu_i(a^n)$ for all n, it follows that $a_i^0 \in \mu_i(a^0)$.

Hence, let $\{a^n\} \subset \prod_{k=1}^h A_k$, $\{a_i^n\} \subset A_i$ be sequences converging to a^0 and a_i^0, respectively, with $a_i^n \in \mu_i(a^n)$ for all n. As $a_i^n \in \gamma_i(a^n)$ for all n and γ_i is upper hemicontinuous, it follows that $a_i^0 \in \gamma_i(a^0)$. Moreover, as γ_i is lower hemicontinuous, for any $z \in \gamma_i(a^0)$ there exists a sequence $\{z^n\} \subset A_i$ converging to z such that $z^n \in \gamma_i(a^n)$ for all n. Thus, $v_i(a_i^n) \geq v_i(z^n)$ for all n, because a_i^n maximizes v_i over $\gamma_i(a^n)$ and, in the limit, $v_i(a_i^0) \geq v_i(z)$. As this inequality holds for every $z \in \gamma_i(a^0)$, we have shown that $a_i^0 \in \mu_i(a^0)$.

We know that μ_i is an upper hemicontinuous correspondence with non-empty, compact and convex values. Therefore, the correspondence $\mu\colon \prod_{k=1}^h A_k \to \prod_{k=1}^h A_k$ given by $\mu(a) = \prod_{i=1}^h \mu_i(a)$ exhibits the same properties. As $\prod_{k=1}^h A_k$ is a non-empty, compact and convex subset of $\mathbb{R}^{lh}$, we can apply Kakutani's fixed point theorem to ensure the existence of some point a^* in $\prod_{k=1}^h A_k$ such that $a^* \in \mu(a^*)$. □

The existence of a social equilibrium, proven originally by Gerard Debreu in 1952, is a neat and powerful result. Yet one has to be careful on interpreting its scope. Because in order to use this theorem, one has to produce reasonable specific models of societies (or economies) whose characteristics induce the properties that its application requires. In other words, we have to ensure that assumptions (A.1) to (A.3) hold, out of the primitives of the model (basic properties concerning the ingredients of the agents choice problems in specific contexts). In particular, there is much more than applying a fixed point argument to show the existence of competitive equilibrium, as we shall see next. The major difficulties are related, not surprisingly, to the continuity of the restrictions and the objective function.

3. From social equilibrium to competitive equilibrium

Let us now consider a general equilibrium model of a competitive economy. Here agents make decisions, relative to goods and services that are traded in the market. The basic elements of the model are: (i) *Commodities and prices*, which are the variables of the problem; (ii) *The agents*, which are the relevant decision units.

We shall consider here, for the sake of simplicity in exposition, the case of a *pure exchange economy*. That is, the only agents are the consumers who trade in the market their possessions. Trade arises out of the diversity in tastes and endowments.

We assume that there is a fixed number l of commodities (a natural number, with $1 \leq l < \infty$). Commodities are goods or services that can be distinguished according to their *characteristics* and their *availability*. The quantity of a commodity will be represented by a real number. This amounts to saying that we assume that commodities are perfectly divisible and implies that we take $\mathbb{R}^l$ as the *commodity space*¶.

Each commodity $h = 1, 2, \ldots, l$ has associated with it a real number p_h representing its price. A *price system* will be represented by a vector $p \in \mathbb{R}^l_+$. Observe that taking p as a point in $\mathbb{R}^l$, which is precisely the commodity space, introduces an implicit assumption which is essential: there is a price for each commodity. This is usually expressed by saying that *markets are complete*.

Economic agents are the decision units of the model. In a pure exchange economy there are only two types of agents: *consumers* and *the Government*. Of these two categories, only the first one will be explicitly modelled, while "the Government" will appear as a central agency that may impose some regulation policies and enforces the property rights.

A consumer is an individual agent (a single household or a family) who takes consumption decisions, that is, decisions referring to the demand for goods and services and the supply of different types of labour. It will be assumed that there is a fixed number m of consumers, indexed by $i = 1, 2, \ldots, m$. The consumer's decision problem is a *problem of choice under restrictions*. The consumer's rational behaviour will be identified with the choice of best options within the set of alternatives that are affordable. There are three elements that define this problem: (a) The *choice set*, that describes the universe of alternatives on which the consumer's choice problem is formulated; (b) The *choice criterion*, that reflects the way in which the consumer evaluates alternative options; and (c) The *restrictions*, that limit the consumer's effective opportunities of choice.

The choice set for consumer i is given by a subset $X_i \subset \mathbb{R}^l$ that describes those consumption vectors that can be realized, given the individual's abilities and biological constraints. A *consumption plan* for the i-th consumer is an l-dimensional vector $x_i \in X_i$. A consumption plan specifies some amounts of goods and labour which the consumer is able to realize. Those goods and services that the consumer demands are usually denoted by positive numbers, whereas her supply of productive factors (different types of

¶Taking $\mathbb{R}^l$ as the commodity space is a convenient assumption, since it exhibits very good operational properties. In particular, it provides both a vector space structure and a suitable topology (e.g., the scalar product is a well defined and continuous operation).

labour) are denoted by negative ones.

The way in which a consumer ranks different consumption plans is described by a binary relation $\succ_i$, called the consumer's *preference relation*, defined as follows: for each pair of points $x_i, x'_i \in X_i$, $x'_i \succ_i x_i$ means that x'_i is *preferred* to x_i. We assume that this preference relation satisfies the following properties:

- Transitivity: $[x'_i \succ_i x_i, x_i \succ_i x''_i] \Rightarrow x'_i \succ_i x''_i$;
- Continuity: the sets $B(x_i) = \{x'_i \in X_i \colon x'_i \succ_i x_i\}$, $W(x_i) = \{x'_i \in X_i \colon x_i \succ_i x'_i\}$ are open in X_i.
- Convexity: $x'_i \succ_i x_i \Rightarrow \lambda x'_i + (1-\lambda)x_i \succ_i x_i$, for all $\lambda \in (0,1)$.
- Monotonicity: $x'_i >> x_i \Rightarrow x'_i \succ_i x_i$.

Transitivity tells us about the consistency in the choice criterion. Continuity establishes that if a point x'_i is better than (resp., worse than) x_i, then points that are close enough to x'_i will also be preferred to x_i. Convexity says that the upper level sets are convex (i.e., intermediate combinations of consumption plans tend to be more appreciated). Finally, monotonicity says that more of all commodities is always better[‖].

In a market economy, the consumer's economic problem consists of choosing a best consumption plan among those that are *affordable* in X_i. Let $p \in \mathbb{R}^l$ be a price vector and $x_i \in X_i$ a consumption plan. Consumer i's expenditure is given by the scalar product $px_i = \sum_{k=1}^{l} p_k x_{ik}$. The i-th consumer's wealth is given by the market worth of her assets; that is, $p\omega_i$, where $\omega_i \in \mathbb{R}^l$ is the i-th consumer's initial endowments. The wealth constraint of consumer i, which determines what is affordable to her, defines her *budget correspondence*, given by a mapping $\beta_i \colon \mathbb{R}^l \to X_i$ with $\beta_i(p) = \{x_i \in X_i \colon px_i \leq p\omega_i\}$.

Observe that β_i is homogeneous of degree zero in p, that is, for every $\lambda > 0$, $\beta_i(\lambda p) = \beta_i(p)$. This means that only relative prices actually matter for consumers' decisions. We can, therefore, substitute the price space by the set $\Delta = \{p \in \mathbb{R}^l_+ \colon \sum_{k=1}^{l} p_k = 1\}$ of *normalized prices* (also called the *price simplex*), a non-empty compact and convex set.

Concerning consumers we shall assume the following[**]:

[‖]This is a crude way of expressing the idea that, in any relevant economic problem, the available commodities are always *scarce* relative to the needs and desires of consumers. Some would postulate that this is the essence of economic problems.

[**]The standard assumption is that the consumption set is bounded from below, which turns out to be a natural assumption in this context. Nothing in the sequel depends on the compactness hypothesis, that will simplify our reasoning.

Axiom 1. *For all $i = 1, 2, \ldots, m$,*

(i) X_i *is a non-empty compact and convex subset of* $\mathbb{R}^l$.
(ii) $\succ_i$ *is a transitive, continuous, convex and monotone preference relation.*
(iii) $\omega_i \in X_i$ *and there exists* $\bar{x}_i \in X_i$ *such that* $\bar{x}_i << \omega_i$.

It is immediate to check that, under Axiom 1, $\beta_i(p)$ is a non-empty, closed convex set for all $p \in \mathbb{R}^l_+$. Part (iii) of Axiom 1 (usually called the "cheaper point" requirement) is a strong assumption that is required in order to ensure that the budget correspondence is continuous and non-empty valued.

The following results are far from trivial (e.g., [6]):

Proposition 3.1. *Under Axiom 1, the preference relation* $\succ_i$ *can be represented by a real–valued function* $u_i \colon X_i \to \mathbb{R}$*, called consumer i's* utility function, *that is continuous, quasiconcave and monotone.*

Proposition 3.2. *Under Axiom 1, the budget correspondence* β_i *is continuous in p, for all* $p \in \Delta$.

Under Axiom 1, in view of Proposition 3.1, the i-th consumer's choice problem can be expressed as the solution to the following program:

$$\left.\begin{array}{l} \max_{x_i} u_i(x_i) \\ \text{s. t.: } x_i \in \beta_i(p) \end{array}\right\} \qquad \text{[P']}$$

The behaviour of consumer i is characterized by the choice of a best affordable option. We define the i-th consumer's *demand correspondence* as a mapping $\xi_i : \Delta \to X_i$ that associates the set of points that solve program [P'] for each price vector.

Observe that since the budget correspondence is homogeneous of degree zero, it follows that $\xi_i(\lambda p) = \xi_i(p)$ for all $\lambda > 0$ (i.e., the demand correspondence inherits the zero homogeneity property of the budget correspondence).

An *allocation* is a point $(x_i)_{i=1}^m$ in the set $\prod_{i=1}^m X_i$. An allocation is a collection of actions, one for each agent, within their respective choice sets (hence a point in $\mathbb{R}^{lm}$). An allocation is *feasible* if $\sum_{i=1}^m x_i \leq \omega$ (i.e., if the aggregate consumption does not exceed the available resources).

Definition 3.1. A *competitive equilibrium* for a pure exchange competitive economy, $E = (X_i, u_i, \omega_i)_{i=1}^m$ is a price vector $p^* \in \Delta$ and an allocation $(x_i^*)_{i=1}^m$ such that:

(a) For all $i = 1, 2, \ldots, m$, $x_i^* \in \xi_i(p^*)$.
(b) $\sum_{i=1}^m x_i^* = \omega$.

A competitive equilibrium is a situation in which all consumers maximize utility within their budget sets at prices p^* and all markets clear.

4. Equilibrium and efficiency

Let us show first that Axiom 1 is sufficient to ensure the existence of a competitive equilibrium as a special case of a social equilibrium (Theorem 2.1). To do so, let $h = m + 1$ with

(a) For all $i = 1, 2, \ldots, m$, $A_i = X_i$, $\gamma_i(a) = \beta_i(p)$, $v_i(a) = u_i(x_i)$.
(b) For $h = m + 1$, $A_h = \gamma_h(a) = \Delta$ for all $a \in \prod_{k=1}^h A_k$ and $v_{m+1}(a) = p\sum_{i=1}^m (x_i - \omega_i)$.

Note that the choice of an individual consumers affects neither others' choice sets nor utility functions (what is usually referred to as "no externalities"). So this is a much simpler setting than that presented above. Only the actions of "the last agent" affect consumers budget sets. Indeed, "the last agent" can be understood as an expression of the functioning of competitive markets (usually identified with *the auctioneer*). Her choice set is the price simplex, and her choice criterion consists of maximizing the worth of the excess demand (in such a way prices go up when demand exceeds supply and viceversa). It is not difficult to prove the following:

Proposition 4.1. *Under Axiom 1, the correspondence* $\pi\colon \prod_{i=1}^m X_i \to \Delta$ *given by*

$$\pi(\cdot) = \arg\max_p \left\{ p\left(\sum_{i=1}^m x_i - \omega\right)\right\}$$

is continuous, with non-empty, compact and convex values.

Propositions 3.1, 3.2 and 4.1 ensure that Axiom 1 induces the necessary conditions to apply Theorem 2.1. Therefore,

Corollary 4.1. *Let E be a pure exchange economy satisfying Axiom 1. Then, a competitive equilibrium exists.*

The existence of equilibrium establishes that competitive markets are institutions that are able to ensure that the exchange process, carried out by price taking agents, can be consistently realized in a decentralized way.

How good is an equilibrium allocation when compared with other feasible alternatives? To answer this question one has to introduce *value judgements*, because different people may rank alternatives differently. Looking for consensus on the way of evaluating allocations, we focus on *the Pareto principle.* This principle says that a feasible allocation is better than another one, when it is preferred by *all* consumers. A Pareto optimum is a maximal element of this relation: there is no feasible allocation in which all agents can be better off.

Asking for Pareto optimality can be regarded as an expression of John Stuart Mill's principle: "the highest welfare for the greatest number". Something really hard to object. The cost of such a broad consensus is that the Pareto principle is not very informative. In particular: (a) Many alternative allocations are not comparable, according to this principle (those in which some consumers are better off and some others are worse off); (b) The set of Pareto optimal allocations can be very large, and include extremely different welfare distributions (this indicates that the Pareto criterion is devoid of any distributive justice feature).

We now show two results that are known as the Two Fundamental Theorems of Welfare Economics. The first one says that every competitive equilibrium is an optimum. The second one establishes that every optimum is an equilibrium, provided that we can freely redistribute wealth among consumers. Hence, if we can select an efficient allocation as a socially desirable outcome, there is a redistribution of initial endowments and a price vector, that yield this particular allocation as a competitive equilibrium. The message of this theorem is twofold: (1) Equity and efficiency are not incompatible aspirations in a competitive economy; (2) The desired outcome can be obtained by a suitable modification of property rights, without having to impose particular actions on individual agents.

Definition 4.1. We say that a feasible allocation $(x_i^0)_{i=1}^m$ is *Pareto optimal* when there is no other feasible allocation $(x_i')_{i=1}^n$ such that, $u_i(x_i^0) \geq u_i(x_i')$ for all i, with $u_i(x_k^0) > u_i(x_k')$ for some k.

Now we can prove:

Theorem 4.1 (First Welfare Theorem). *Let E be a pure exchange competitive economy in which every consumer has a locally non-satiated utility function. Let $(p^*, (x_i^*)_{i=1}^m)$ be a competitive equilibrium. Then the allocation $(x_i^*)_{i=1}^m$ is Pareto efficient.*

Proof. Suppose that this is not true, that is, there exists a feasible allocation $(x'_i)_{i=1}^m$ such that $u_i(x'_i) \geq u_i(x^*_i)$ for all i, with $u_i(x'_k) > u_i(x^*_k)$ for some k. Monotonicity implies that $p^*x'_i \geq p^*x^*_i$ for all i, with $p^*x'_k > p^*x^*_k$ for those consumers with $u_k(x'_k) > u_k(x^*_k)$. Hence, $p^* \sum_{i=1}^m x'_i > p^* \sum_{i=1}^m x^*_i$. As $p^*x^*_i = p^*\omega_i$, this in turn implies: $p^* \sum_{i=1}^m x'_i > p^* \sum_{i=1}^m \omega_i$. Moreover, $\sum_{i=1}^m x'_i \leq \omega$ because $(x'_i)_{i=1}^m$ is a feasible allocation. Therefore, $p^* \sum_{i=1}^m x'_i \leq p^* \sum_{i=1}^m \omega_i$, against the former conclusion. □

Theorem 4.2 (Second Welfare Theorem). *Let $E = [(X_i, u_i), \omega]$ be an economy that satisfies Axiom 1 and let $(x^*_i)_{i=1}^m$ be a Pareto optimal allocation, with $x_i \in^*$ int X_i for all i. Then, there exists a price vector $p^* \in \Delta$ and a wealth distribution such that $(p^*, (x^*_i)_{i=1}^m)$ is a competitive equilibrium.*

Proof. Let $x = \sum_{i=1}^m x_i$, $X = \sum_{i=1}^m X_i$. Define $\mathrm{BE}(x^*) \equiv \{x \in X \colon u_i(x_i \geq u_i(x^*_i), \forall i\}$, a convex set. As $(x^*_i)_{i=1}^m$ is feasible, $x^* \leq \omega$, it follows that $\omega \in \mathrm{BE}\ (x^*)$. Moreover, $\omega \notin \mathrm{int\ BE}(x^*)$ because $(x^*_i)_{i=1}^m$ is a Pareto optimal allocation. Therefore, we can find a vector $p^* \neq 0$ such that $p^*\omega \leq p^*x$ for all $x \in \mathrm{BE}(x^*)$. This implies $p^*x^* = \min p^*x$ for all $x \in \mathrm{BE}(x^*)$ with $p^*x^* = p^*\omega$. Hence, x^* minimizes aggregate expenditure at prices p^* on $\mathrm{BE}(x^*)$. It is easy to see that this implies that x^*_i is an expenditure minimizing consumption plan at prices p^* on the set of consumption plans that are better than or equal to x^*_i for all i. Thus, x^*_i is the i-th consumer's demand at prices p^*.

So if we let $\omega_i = x^*_i$, $i = 1, 2, \ldots, m$, it follows that $[p^*, (x^*_i)_{i=1}^m]$ is a competitive equilibrium for this economy. □

5. Final comments

We have presented here the basic ingredients of a general equilibrium model of a competitive economy, in a highly simplified scenario. The model, however, is robust enough to apply to much richer environments. Let us conclude by briefly commenting on those extensions.

The model presented here refers to the special case in which no production activities take place (a "pure exchange" economy). This is not a relevant restriction as long as we keep the competitive scenario. Indeed, competitive firms are easily accommodated into the model (they can be regarded as just providing a continuous transformation of the available resources). The case in which firms do really matter is that in which markets are not competitive. Yet we are far from solving a full fledged general equilibrium model with non-competitive firms.

We have assumed compact choice sets, monotone preferences, and absence of externalities. None of those assumptions are necessary to ensure the existence of equilibrium (even the transitivity of the preference relation can be dispensed with). There are also equilibrium models with a continuum of agents or an infinite number of commodities.

Even though the externalities can be introduced in the model without affecting the existence of equilibrium (indeed the social equilibrium adopts such an approach), they kill the efficiency properties. The reason is that agents make decisions regarding their private interests without paying attention to the effects those actions produce on other agents. The standard setting in which those externalities play a role is that of contamination, public goods or resources of common property.

The convexity assumption (convex choice sets and quasiconcave objective functions) is, however, hard to avoid. The same applies to the conditions that ensure the continuity of the mappings that describe the behaviour of the agents.

Finally, let us point out that the existence of equilibrium tells us about the possibility that markets are able to coordinate the economic activity. Yet from this result one cannot deduce that market forces *drive* the economy towards an equilibrium. Indeed, we know rather well *when* the existence of an equilibrium can be ensured, and know very little about *how* this happens (if indeed it happens) and how fast it does.

The classic works of Arrow and Debreu ([1]) and McKenzie ([10]) are still exciting readings. Debreu ([6,7]), Arrow and Hahn ([2]), Cornwall ([4]) or Mas–Colell, Whinston and Green ([9]), among many others, discuss this problem more thoroughly. Debreu ([5]) offers a number of interesting extensions to the model presented here. Border ([3]) offers a nice summary of the results and the techniques involved. Villar ([12]) provides an analysis of a general equilibrium model with non-convex firms.

There are three classical topics in the analysis of equilibrium that we have omitted: the uniqueness, stability and core properties of competitive equilibria. Mas–Colell *et al.* ([9]) provide a suitable introduction to the analysis of these topics. Hildenbrand and Kirman ([8]) and Cornwall ([4]) contain detailed analysis of the core and its connection with competitive equilibria. Further references can be found there.

References

1. K. J. Arrow and G. Debreu, *Existence of Equilibrium for a Competitive Economy*, Econometrica **22** (1954), 265–290.

2. K. Arrow and F. Hahn, *General Competitive Analysis*, Holden Day, San Francisco (1971).
3. K. C. Border, *Fixpoint Theorems with Applications to Economics and Game Theory*, Cambridge University Press, New York (1983).
4. R. E. Cornwall, *Introduction to the Use of General Equilibrium Analysis*, North Holland, Amsterdam (1984).
5. G. Debreu, *A Social Equilibrium Existence Theorem*, Proceedings of the National Academy of Sciences **38** (1952), 886–893.
6. G. Debreu, *Theory of Value*, Wiley and Sons, New York (1959).
7. G. Debreu, *Existence of Competitive Equilibrium*, Arrow and Intriligator, chapter 15 (1982).
8. W. Hildenbrand and A. Kirman, *Introduction to Equilibrium Analysis*, North Holland, Amsterdam (1988).
9. A. Mas–Colell, M. D. Whinston and J. R. Green, *Microeconomic Theory*, Oxford University Press, New York (1995).
10. L. McKnezie, *On the Existence of General Equlibrium for a Competitive Market*, Econometrica **27** (1959), 54–71.
11. A. Smith, *An Inquiry into the Nature and Causes of the Wealth of Nations* (1776).
12. A. Villar, *Equilibrium and Efficiency in Production Economies*, Springer–Verlag, Berlin (2000).
13. L. Walras, *lments d'conomie politique pure, ou thorie de la richesse sociale* (1874).

IDEALS IN F-ALGEBRAS

WIESLAW ŻELAZKO

Mathematical Institute, Polish Academy of Sciences
Śniadeckich 8, P.O. Box 21
00-956 Warsaw, Poland
E-mail: zelazko@impan.gov.pl

We give here a survey on several open problems and several related, mostly recent, results concerning ideals in F-algebras and its subclasses (B_0-algebras and m-convex B_0-algebras).

1. Introduction

F-algebras are topological algebras (i.e., topological vector spaces equipped with a jointly continuous associative multiplication), which are F-spaces, i.e., complete metric topological vector spaces. On one hand, it is quite large class containing Banach algebras, locally bounded algebras (a class of topological algebras containing Banach algebras; we shall not deal here with this class, since the properties of their ideals are essentially the same as those for Banach algebras) and B_0-algebras. On the other, hand it is narrow enough in order to obtain there interesting results. This paper is a survey containing some general results concerning ideals in F-algebras and stating or recalling some open problems important for further development of the theory of topological algebras (note that an arbitrary complete topological algebra is an inverse limit of F-algebras [12]). For sake of simplicity and avoiding unnecessary complication, we shall be considering only unital (real or complex) algebras. For an information about general topological algebras, the reader is referred to Mallios [8], see also [16].

2. Prerequisities

The concept of a F-space (Fréchet space) was introduced by Banach ([5]). Later, French mathematicians added there the condition of local convexity; such spaces are called B_0-spaces by the Polish school (see [9] and [10]) and we shall keep here this terminology. The topology of a F-space X can be given by means of an F-norm, i.e., a functional $\|\cdot\|$ on X satisfying the following conditions:

(i) $\|x\| \geq 0$ and $\|x\| = 0$ iff $x = 0$,
(ii) $\|x + y\| \leq \|x\| + \|y\|$,
(iii) the map $(\lambda, x) \mapsto \lambda x$ is jointly continuous.

Here x, y are elements in X and λ belongs to real or complex scalars. A distance giving the topology of X can be given by $\|x - y\|$, and X is assumed to be complete with respect to this distance (then it is also complete with respect to all equivalent F-norms, i.e., giving the same topology on X). For more information about F-spaces the reader is referred to [5] and [13].

An F-algebra A is a F-space equipped with an associative jointly continuous multiplication making it a complex or real algebra. However, there is no particular relation between F-norms $\|x\|, \|y\|$ and $\|xy\|$ expressing the joint continuity of multiplication. The class of all F-algebras will be denoted by $\mathcal{F}$.

The topology of a B_0-space X can be given by means of an increasing sequence of homogeneous seminorms

$$\|x\|_1 \leq \|x\|_2 \leq \ldots, \qquad x \in X, \tag{1}$$

and $\lim_i x_i = y$ if $\|x_i - y\|_k \to 0$ for all natural k. An F-norm giving the topology of a B_0-space can be given by

$$\|x\| = \sum_{k=1}^{\infty} 2^{-k} \frac{\|x\|_k}{1 + \|x\|_k}.$$

If A is a B_0-algebra, then its seminorms (1) can be chosen so that

$$\|xy\|_k \leq \|x\|_{k+1} \|y\|_{k+1}, x, y \in A, \quad \text{and} \quad \|e\|_k = 1, \quad k = 1, 2, \ldots, \tag{2}$$

where e is the unity of A (see [15] or [16] and [7]). The class of all B_0-algebras will be denoted by $\mathcal{B}_0$.

For some algebras it is possible to choose seminorms (1) so that (2) can be replaced by

$$\|xy\|_k \leq \|x\|_k \|y\|_k, \quad \text{and} \quad \|e\|_k = 1. \tag{3}$$

Such B_0-algebras are called multiplicatively convex (shortly m-convex) and the class of all m-convex B_0-algebras will be denoted by $\mathcal{M}B_0$.

Finally, the class of all Banach algebras will be denoted by $\mathcal{B}$. clearly, we have

$$\mathcal{B} \subset MB_0 \subset \mathcal{B}_0 \subset \mathcal{F}.$$

We shall provide now some examples of non-Banach F-algebras which shall be useful in the sequel.

Example 2.1. Denote by $\mathcal{E}$ the algebra of all entire functions of one complex variable provided with the compact–open topology (the topology of uniform convergence on compact subsets of the complex plane). This topology can be given by seminorms

$$\|x\|_k = \max_{|\zeta|\leq k} |z(\zeta)|.$$

Clearly these seminorms satisfy (1) and (3), and so $\mathcal{E}$ is an m-convex B_0-algebra.

Example 2.2. Denote by $C^\infty[0,1]$ the algebra of all infinitely derivable functions with all derivatives continuous on the closed unit interval, and provided with the topology of uniform convergence of functions together with all their derivatives. This topology can be given by seminorms

$$\|x\|_k = 2^k \max_{0\leq i\leq k} \max_{0\leq t\leq 1} 2^i |x^{(i)}(t)|.$$

It is not difficult to verify that these seminorms satisfy (1) and (3), and so the algebra of this example is an m-convex B_0-algebra.

Example 2.3. Denote by (s) the algebra of all formal power series $x = \sum_{k=0}^{\infty} \xi_k(x)t^k$ with the topology of pointwise convergence of the coefficients $\xi_k(x)$ and with the Cauchy multiplication of power series. It is again an m-convex B_0-algebra with seminorms

$$\|x\|_k = \sum_{i=0}^{k-1} |\xi_i(x)|,$$

which satisfy (1) and (3).

The following non-commutative version, constructed in [20], of the algebra (s) will be useful in Section 4. Let w be an additional variable and define the algebra A as the direct sum $A = (s) + (s)w$ with the multiplication given by the relations $w^2 = 0, wt = 0$, so that the product of two elements in A is given by the formula

$$(a + bw)(c + dw) = ac + (ad + \xi_0(c)b)w, \quad a, b, c, d \in (s). \tag{4}$$

The topology of A is given by the seminorms $|a + bw|_n = \|a\|_n + \|b\|_n, n = 1, 2, \ldots$. It is easy to see that A is again an m-convex B_0-algebra.

Example 2.4. Put $L^\omega[0,1] = \cap_{1\le p<\infty} L_p[0,1]$, with the pointwise algebra operations. The L_p spaces are taken with respect to the Lebesgue measure on the unit interval and the topology of L^ω is given by convergence in all L_p norms $\|\cdot\|_p$. The same topology is given by any sequence p_n tending to an infinity. Choosing $p_n = 2^{n-1}$ we obtain a sequence of seminorms satisfying relations (1) and (2). Thus L^ω is a B_0-algebra and it can be shown (see e.g. [15] or [16]) that it is not m-convex.

Example 2.5. Let $(a_{n,k})$ be a matrix of positive real numbers, $n = 1, 2, \ldots$, and k runs either over all integers or over all non-negative integers, such that

$$a_{n,k+l} \le a_{n+1,k} a_{n+1,l} \quad \text{and} \quad a_{n,0} = 1, \tag{5}$$

for all involved n, k, l. The matrix algebra $M(a_{n,k})$ consists of all formal series of the form

$$x = \sum_k \xi_k(x) t^k$$

which are Laurent series if k runs over all integers, or power series, if k runs over all non-negative integers. We assume that

$$\|x\|_n = \sum_k a_{n,k} |\xi_k(x)| < \infty, \quad n = 1, 2, \ldots, \tag{6}$$

and the multiplication is Cauchy multiplication of Laurent or power series. The conditions in (5) imply that the seminorms given by (6) satisfy (1) and (2), so that matrix algebra are B_0-algebras, which can be also m-convex. Many interesting examples can be obtained in this way. We mention here the Williamson's algebra ([14], see also [15] or [16]) given by the matrix

$$a_{n,k} = \begin{cases} (1+k)^{-\frac{1+k}{n}} & \text{for } k \ge 1 \\ 1 & \text{for } k = 0 \\ (1-k)^{n(1-k)} & \text{for } k \le -1. \end{cases}$$

It is a non-m-convex B_0-algebra. The Williamson's algebra contains a dense subalgebra isomorphic with the field of rational functions in one variable.

Example 2.6. Denote by $L_0[0,1]$ the algebra of all Lebesgue measurable (almost everywhere finite) functions with pointwise algebra operations and with the topology of convergence in measure. This topology can be given by the F-norm given by the formula

$$\|x\| = \int_0^1 \frac{|x(t)|}{1 + |x(t)|}\, dt.$$

This algebra is not locally convex.

3. Basic properties of F-algebras

The main properties of Banach algebras on which it is based their theory are

(i) The group $G(A)$ of invertible elements of a Banach algebra A is open (a topological algebra with this property is called a Q-algebra).
(ii) The operation of taking an inverse $x \mapsto x^{-1}$ is continuous on $G(A)$.
(iii) The Gelfand–Mazur theorem holds true for Banach algebras, i.e., every complex Banach division algebra is isomorphic to **C** and every real Banach division algebra is isomprphic to either of **C, R, H**, where **H** denotes the division algebra of quaternions.

The F-algebras with the property (i) are rather rare. In the above examples, the algebra $C^\infty[0,1]$ of Example 2.2, and the algebras (s) and A of Example 2.3 enjoy this property, and the algebra $\mathcal{E}$ of Example 2.1 has, in a certain sense, an opposite property: the set $G(A) \cup \{0\}$ is closed.

The property (ii) holds true for all m-convex algebra and also for $L_0[0,1]$ (Example 2.6) but fails for the algebra L^ω of Example 2.4 and for the Williamson's algebra of Example 2.5. For all F-algebras, the property (i) implies the property (ii) as follows from the following result (see [14] or [15]).

Theorem 3.1. *Let A be a real or complex F-algebra. Then the operation $x \mapsto x^{-1}$ is continuous on $G(A)$ if and only it is a G_δ-set.*

In particular, for all m-convex B_0-algebras, $G(A)$ is a G_δ-set. In non-metrizable topological algebras Theorem 3.1 fails to be true: there are such algebras with a continuous inverse for which $G(A)$ is not a G_δ)-set and there are algebras with discontinuous inverse for which $G(A)$ is even open.

The Gelfand–Mazur theorem holds true for all m-convex algebras and for all B_0-algebras (see [11], [15] and [16]). In general, it fails to be true for non-metrizable algebras, even for locally convex ones. The following problem remains open for many years (see [15] and [16]).

Problem 3.1. *Is the Gelfand–Mazur theorem true for F-algebras?*

In the context of F-algebras we meet another concept of invertibility, namely the *topological invertibility.* We call an element x of an F-algebra A left topologically invertible if there is a sequence $(z_n) \subset A$ such that $\lim_n z_n x = e$, where e is the unity of A. Similarly we define right topologically invertible elements. An element $x \in A$ is topologically invertible if it is both left and right topologically invertible. A topologically invertible element is said proper if it is non-invertible. The m-convex B_0-algebras do not have proper topologically invertible elements, but those elements do exist in $L^\omega[0,1]$ (the function $x(t) = t$ is such an element) and also in the Williamson's algebra. The following result is given in [18] (it is proved there for commutative algebras, but the proof works also in the non-commutative case).

Theorem 3.2. *Let A be a F-algebra, then the set $G^t(A)$ of all its topologically invertible elements is a G_δ-set.*

So, in view of theorem 3.1, every F-algebra with a discontinuous inverse must possess proper topologically invertible elements. We do not know whether the converse is true.

Problem 3.2. *Suppose that a F-algebra has a proper topologically invertible element. Does it follow that the inverse is discontinuous on $G(A)$?*

The answer is not known even in the commutative case.

For more information about topological algebras, and for more examples, the reader is referred to [8], [11], [15] and [16].

We pass now to the main topic of this paper.

4. Ideals in F-algebras

First we shall consider the following question.

Question 4.1. Let A be a real or complex F-algebra. When all maximal ideals in A are closed?

Here we have in mind the left, right, and two–sided ideals. For left or right ideals the complete answer is given by the following result in [21] (recall that A is a Q-algebra if the set $G(A)$ is open).

Theorem 4.1. *Let A be a unital F-algebra. Then the following conditions are equivalent.*

(i) All maximal left ideals in A are closed.
(ii) The set $G_l(A)$ of all left–invertible elements in A is open.
(iii) A is a Q-algebra.
(iv) The set $G_r(A)$ of all right–invertible elements in A is open.

(v) *All maximal right ideals in A are closed.*

For non-metrizable topological algebras the above result fails to be true. The above result implies that in the class of F-algebras there cannot exist an algebra for which the set $G_l(A)$ is open, but which is not a Q-algebra, i.e., there does not exist a proper O_l-algebra. But we do not know whether such an algebra can exist for more general topological algebras. We do not have a satisfactory answer to the Question 4.1 in the case of a two–sided ideal. In [17], we defined Q_2-algebras as (unital) topological algebras A for which the set $G_2(A)$ is open. This set is defined as the set of all elements x in A for which the smallest two–sided ideal generated by x coincided with the whole of A. The result obtained in [17] says that and F-algebra A has all maximal two–sided ideals closed if and only if A is a Q_2-algebra. But we do not know whether (non-commutative) Q_2-algebras coincide with Q-algebras. If such a coincidence holds true, then a F-algebra topology for a simple algebra (an algebra for which the only two–sided ideal is the zero ideal) is always a Q-algebra topology.

For the formulation of the next result we need following definitions. We say that an ideal is finitely generated by elements $x_1, \ldots, x_n$ if it coincides with the smallest ideal containing these elements (no closure is taken here). Let A be a unital topological algebra and $m(A)$ the family of all its closed maximal two–sided ideals. The hull–kernel topology on $m(A)$ is given by the means of a closure operation: for $S \subset m(A)$, define its closure as $hk(S)$, where $k(S)$ is the intersection of all ideals in S, and $k(I)$, for a two–sided ideal I in A, is the set of all ideals in $m(A)$ containing I. The hull–kernel topology is, generally speaking, non-Hausdorff.

The following result, valid for all topological algebras, is due to Abel and Jarosz ([1, theorem 1]).

Theorem 4.2. *Let A be a unital topological algebra. Then all its maximal two–sided ideals are closed if and only if*

(i) *Every proper finitely generated two–sided ideal of A is contained in some ideal belonging to $m(A)$, and*
(ii) *$m(A)$ is compact in the hull–kernel topology.*

Again we do not know what this result means in the case when A is a F-algebra. The condition (ii) may suggest that closedness of all maximal two–sided ideals of A is equivalent with the fact that A is a Q-algebra. Also the condition (i) adds some importance to our further Question 4.3.

Theorem 4.1 implies that any F algebra which is not a Q-algebra must have a dense right maximal ideal and a dense left maximal ideal (and a dense maximal ideal in the commutative case). So the algebras of Examples 2.1, 2.4, 2.6 and the Williammson's algebra of Example 2.5 have dense maximal ideals and all of them are of infinite codimension. In fact, all maximal ideals in L^ω, $L_0[0,1]$ and in the Williamson's algebra are dense. Every commutative m-convex complex (unital) B_0-algebra must have a continuous multiplicative–linear functional (character), and its kernel is a closed maximal ideal. The Gelfand–Mazur theorem for B_0-algebras implies that in such a commutative algebra every closed maximal ideal is kernel of a character, but such a fact is not known for general F-algebras (cf. Problem 3.1). The following problem is still open.

Problem 4.1. *Let A be a F-algebra. Is every character of A continuous?*

In the particular case A is an m-convex B_0-algebra, it is the famous Michael–Mazur Problem (cf. [11,15,16]). Call a topological algebra *finitely generated* by elements $x_1, \ldots, x_n$ if it coincides with its smallest closed subalgebra containing these elements. For finitely generated MB_0-algebras, Arens [3] solved this problem in positive, but for F-algebras or B_0-algebras it is open even for singly generated algebras.

Question 4.2. Let A be a F-algebra. When are all left (right) ideals closed?

Such a question makes a sense since, contrary to the result of Theorem 4.1, it is possible to have an algebra with all left, but not all right ideals closed. Consider the algebra (s) of the Example 2.3. It is not hard to see that every ideal of this algebra is of the form $I_n = t^n(s)$ (it follows from the fact that an element x in (s) is invertible if and only if $\xi_0(x) \neq 0$). Thus, (s) is Noetherian, i.e., every strictly increasing chain of ideals is finite. The algebra A of Example 2.3 is left Noetherian, but not right Noetherian, and all its left ideals are closed while there are non-closed right ideals (e.g. any vector subspace of $(s)w \subset A$ is a right ideal and we can have such a non-closed subspace). For details concerning this example, see [20].

Problem 4.2. *Is it true that all left (resp., right) ideals of a F-algebra A are closed if and only if A is left (resp., right) Noetherian?*

It is quite easy to see that if a F-algebra A has all left ideals closed, then it is left Noetherian, and the problem lies in proving the converse implication. For m-convex B_0-algebras, a positive answer for the Problem 4.2 was given by Choukri, El Kinani and Oudadess ([6]). An answer to a weaker problem, also supporting the conjecture that the answer to Problem 4.2 should be in positive, is given in the following result ([19]).

Theorem 4.3. *Let A be a unital real or complex F-algebra. Then A has all one–sided ideals closed if and only if it is both left and right Noetherian.*

Of course, if A has all left ideals closed, then it has also all two–sided ideals closed. In the proof of the above result and similar results, there are involved topologically invertible elements.

Question 4.3. When a F-algebra has a dense finitely generated ideal?

Clearly, a F-algebra A has a dense principal (singly generated) left ideal if and only if it has a proper topologically left invertible element x (and then the principal ideal Ax is dense). As it was mentioned earlier, such elements and so such dense ideals exist in the commutative algebras $L^{\omega}[0,1]$ and in the Williamson's algebra. On the other hand, Arens has shown in [3] that m-convex algebras cannot have dense finitely generated one–sided ideals (the result was formulated for commutative algebras but, as was observed in [5], its proof works also in the non-commutative case). It can be also shown that the algebra $L_0[0,1]$ has no dense finitely generated ideals. The real problem (for F- or B_0-algebras) is whether the existence of a finitely generated one–sided dense ideal implies the existence of a singly generated one, i.e., the existence of a one–sided topologically invertible element. The situation is unclear for two–sided ideals. Does the existence of such a singly generated ideal imply the existence of a one–sided or two–sided topologically invertible element (the inverse implication is obvious)?

Question 4.4. When a F algebra has all non-zero ideals dense?

It is a long lasting and very famous problem, whether every non-unital Banach algebra must have a proper closed (one–sided, two–sided) ideal. The problem is open also for commutative algebras. However, our question concerns unital algebras. Closed proper ideals must always exist in unital m-convex algebras. Aharon Atzmon, however, constructed an infinite dimensional commutative, complete, locally convex algebra in which all non-zero ideals are dense ([4]). The problem of existence of such an algebra is open for F- or B_0-algebras. So we pose

Problem 4.3. *Does there exist an infinitely dimensional commutative unital F-algebra without proper closed ideals?*

Formally speaking, if there is an infinite dimensional field of type F, then it is such an algebra, but the question is about algebras not being a field. The algebras of Examples 2.1, 2.2, 2.3 and 2.5 have proper closed ideals. Also many matrix algebras (Example 2.4) have such ideals. In many cases there exist even maximal closed ideals. The situation is unclear for the Williamson's algebra. The author does not know any proper (i.e., different from the zero ideal and the whole algebra) closed ideal in this algebra. One can suspect that the Williamson's algebra has no proper closed ideals. The completeness is essential here since the Williamson's algebra contains the field of rational functions which has no proper closed ideal.

References

1. M. Abel and K. Jarosz, *Topological algebras in which all maximal two–sided ideals are closed*, Banach Centre Publ. **67** (2005), 35–43.
2. R. F. Arens, *The space L^ω and convex topological rings*, Bull. Amer. Math. Soc. **52** (1946), 931–935.
3. R. F. Arens, *Dense inverse limit rings*, Michigan Math. J. **5** (1958), 169–182.
4. A. Atzmon, *An operator without invariant subspaces on a nuclear Fréchet space*, Ann. of Math. **117** (1983), 669–694.
5. S. Banach, *Théorie des Opérations Linéaires*, Warszawa (1932).
6. R. Choukri, A. El Kinani and M. Oudadess, *Algèbres topologiques à idéaux à gauche fermés*, Studia Math. **168** (2005), 159–164.
7. A. Fernández and V. Müller, *Renormalizations of Banach and locally convex algebras*, Studia Math. **96** (1990), 237–242.
8. A. Mallios, *Topological Algebras*, Selected Topics, North Holland (1986).
9. S. Mazur and W. Orlicz, *Sur les espaces metriques linéaires I*, Studia Math. **10** (1948), 184–208.
10. S. Mazur and W. Orlicz, *Sur les espaces metriques linéaires II*, Studia Math. **13** (1953), 137–179.
11. E. Michael, *Locally Multiplicatively–Convex Topological Algebras*, Mem. Amer. Math. Soc. **11** (1952).
12. T. Müldner, *Projective limits of topological algebras*, Coll. Math. **33** (1975), 291–294.
13. S. Rolewicz, *Metric Linear Spaces*, PWN, Warszawa (1982).
14. J. H. Williamson, *On topologizing of the field C(t)*, Proc. Amer. Math. Soc. **5** (1954), 729–734.
15. W. Żelazko, *Metric Generalizations of Banach Algebras*, Dissertationes Math. (Rozprawy Mat.) **47** (1965).
16. W. Żelazko, *Selected Topics in Topological Algebras*, Aarhus Univ. Lecture Notes Series **31** (1971).
17. W. Żelazko, *Some problems concering Q-algebras*, Publ. Ecole Norm. Sup. Takaddoum, Non-Normed Topological Algebras, 74–79, Rabat (2004).
18. W. Żelazko, *When a commutative F-algebra has a dense principal ideal*, Contemporary Math. Vol. 341, Providence 2004.

19. W. Żelazko, *A characterization of F-algebras with all one–sided ideals closed*, Studia Math. **168** (2005), 135–145.
20. W. Żelazko, *An m-convex B_0-algebra with all left but not all right ideals closed*, Coll. Math. **104** (2006), 317–324.
21. W. Żelazko, *When a unital F-algebra has all maximal left (right) ideals closed?*, Studia Math. **175** (2006), 279–284.